Romance of Geology

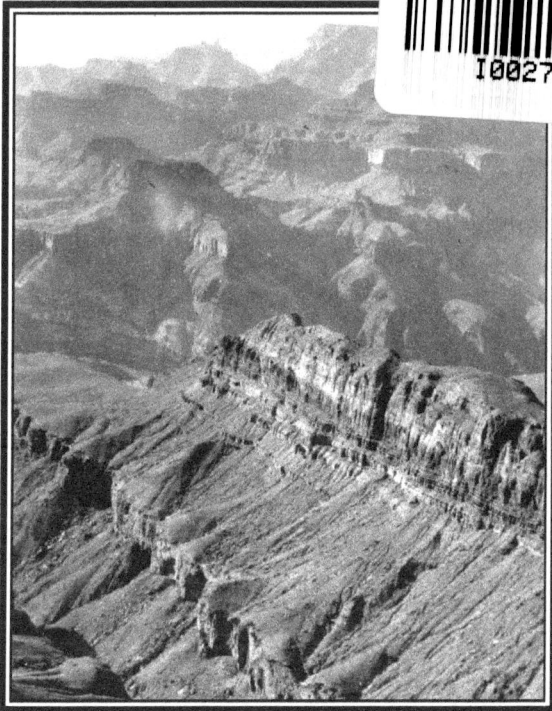

With best wishes of

Enos A. Mills

Temporal Mechanical Press
Long's Peak, Colorado

Temporal Mechanical Press
a division of Enos Mills Cabin
6760 Highway 7
Estes Park, CO 80517-6404
www.enosmills.com
info@enosmills.com

ISBN 978-1-28878-46-9

Title page photograph, the Grand Canyon, by Enos A.
Mills.

Cover photographs and design © Eryn V. Mills. Yardangs
in the Western Desert of Egypt.

Contents

"Watch Your Step!"

Foreword

Everyone should have a hobby—an outside interest. Nature supplies materials for thought and pleasantly compels thinking. We see what we are trained to see, and association with nature develops the faculty of thorough, all-seeing observation. The study and enjoyment of geology and glaciology, observation of the manners and customs of the beaver and the bear, give physical, mental, and sympathetic development of the best possible kind.

Playing in the out-of-doors, on peak or plain, among trees and waterfalls, calms and storms, is one of the surest ways of training the senses and keeping the matchless torch of imagination ever burning brightly.

Scenery, like music, is thought compelling; it shakes us free from ourselves and the world. The crumbling cliff, the glacial landscape, the wild free clouds, the stars and the sky, the white, white cascades, the colored flowers, the untrimmed and steadfast trees, the shadows on the ground, the tangled grass, the round, sunny hills—all these arouse thought, wonder, and delight.

It has seemed to me most unfortunate that people of today do not have firsthand acquaintance with the elements, and I have made it my chief aim in life to arouse interest in the outdoors. During my years of guiding, I added a touch of romance, adventure, and poetry to the wonderful stories which nature—through the oft-changing and evolving conditions of ages—had given to flower and tree, bird and animal, to seed, to soil, and to rusty, rocky pinnacle, wasting through the season, holding its brief place on Time's horizon.

I featured the fact that the wilderness is friendly; wild animals are not ferocious, the out-of-doors is helpful at all times, and recreation is a necessity. A correct knowledge of natural history, of our natural resources, and of the world in which we live is the most important part of education—of right thinking.

Camping in every state in the Union has been my good

fortune, my fun, and my schooling. Alone and unarmed, I have visited the silent places in the snows and flowers. I am glad to have had so many storms and stars, so many days and nights out with nature. It is good to have gone up into the sky on the broken peaks; to have dared the narrow unbanistered ledges and ignored the threat of desert wastes. I have had many a magic night alone in the silence watching my campfire—watching shadows shift and dance upon the cliffs; have lain back upon the balsam boughs and peered into the starlight vault between the tall and slender spires of the spruces, while around me in the solemn forest the coyotes gave their lonely call or weird answer in the depths of the night. Many a dark night I have seen dissolve and develop into all the pictured world of light and shade and beauty of the day. Many a morning sunburst I have beheld glorify the steep and snowy heights. When only a boy, I heard nature's bugle song on Alaska's scenic shores. Intense and happy days I have had with only bark, berries, and mushrooms to eat. But what did eating matter! I felt the occult eloquence of the tongueless scenes; the world was young. Life with nature is always real, and sometimes, with storm or snow slide, it is in deadly earnest.

I cannot transfer to you the rapture that nature has within me stirred, but you are welcome to go out into the vast cathedral of the world and hear the melody blown by the mystic trumpeter, where your eager heart will feel the olden, golden roundelay. "The pathway to the Heroic Age out with Nature lies."

Enos A. Mills.

Imagination Guides Our Race

During the long centuries between cave and cottage our good ancestors traveled Nature's inspiring pictured scenes. With interest and with awe they watched the silent movements of the clouds across the sky, they listened with speechless wonder to the mysterious, unseen echo that lived and mimicked in the air, they puzzled over the strange, invisible wind that shook the excited trees and whispered in the rustling grass. They heard the echoing crash of thunder, saw lightning's golden rivers in the cloud mountains and looked with childish joy upon the silken rainbow. They marveled at the wondrous sunrise, the light of day, the fireflies in the forest, and the lonely, changing moon. The mysterious darkness was never understood but the silent, faithful stars they named and watched with nightly wonder. By trail and campfire these thought-filled wonders took life and color, became poetic stories. Through the changing seasons and the passing years Nature built the brain and kindled the illuminating imagination—the immortal torch that guided our advancing race and which triumphantly leads us on.

Introduction

Enos A. Mills was a man that could not only see the Divine Creator in a blade of grass, but in everything created by that entity. He, like many others of rare insight, saw the science of Nature as The Creator's communication to us detailing the continuous Creation. From his Quaker background was the basic belief that you do not harm what you respect. It is a natural conclusion that if you respect The Creator, you do not harm the Creation, an equal gift, to be honored, appreciated and shared. To Enos, Nature was the Supreme Being's most sacred church.

Fortunately for us, Enos did more than honor and respect all of Nature, he was highly magnetic with his joyful exuberance for attracting people to simply taking the time to look around them, reflecting his personal spiritual beliefs, not preaching them.

In studying geology, one remembers just how artificial clocks, calendars, money, power and laws created by humans are, and quite illusionary in the midst of in Nature, where the laws of the Divine Physicist prevail. We benefit from Enos' experiences in this book to remind us how fleeting our time here is and that it is our own personal choice of traversing through it with integrity and honor, peacefully.

Dear Reader, may your journey through life be blessed with many interesting questions that make life rich and wondrous.

Sight-Seeing by Wireless

Desert mirages are ever on exhibition in the land of little rain. I was riding an Indian Pony along the margin of the Black Rock Desert in Nevada. The afternoon had been filled with a series of tantalizing mirages. A lake, a grove, and green fields had again and again been shown to the right of us and to the left of us. Canteens were empty. All concentration for a spring, I ceased to notice the mocking pictures of the mirage. We were seriously wondering if we had made a mistake in failing to turn off to search for water in the mouth of that desolate canyon. The map placed a spring here, but finding a spring where not a tree stood and in a locality without distinguishing landmarks would be a case mostly of luck.

My pony with a sudden stop gave welcoming neigh. His eager ears pointed to two loose horses less than a quarter of a mile off. A bay and a pinto were standing in alkaline sand near the mouth of a canyon. They suggested camp and a spring. All concentration, my pony started gingerly for them. Head to head the horses stood; possibly looking at each other, possibly staring sleepily at the hot dull sand. I looked through my glass farther into the gorge beyond them. My pony made another sudden stop. The horses vanished.

"Well, Piute," I said, "I am a tenderfoot in the Great Basin, but you—born on the desert—have let a mirage deceive you."

He looked this way and that. He did not act foolish; he still had faith in his eyes and those horses. They simply had dodged while he was hustling over the high sand dunes. He wanted to look behind a rock ledge and I let him; then he clambered over the dunes. Unwillingly, he turned back into our course along the dim trail. After going a quarter of a mile, he turned, looked, neighed wildly, listened, then—puzzled—went slowly forward.

The wireless transfers sounds and music through uncharted space. On a desert, light becomes wireless—

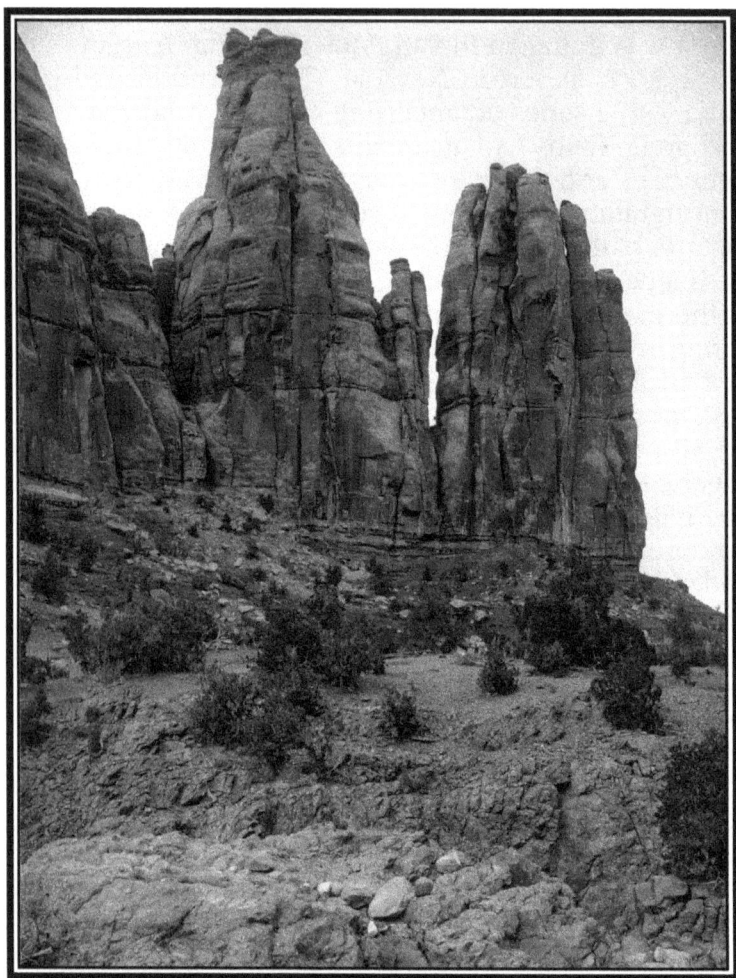

Colorado National Monument.
Photo by George L. Beam.

transfers or transplants scenes. These may be brought from beyond the horizon and placed in the foreground in the vision of the beholder. It often is impossible to tell how much of the scenic desert is real and how much is mirage—scenes transferred by wireless.

We were on the edge of an old lake bottom as level as a floor, an extensive plain, or basin, surrounded by detached mountain ranges. A range on our left rose several thousand feet. It was so narrow and so abrupt, there was a suggestion that its foothills were buried. Earlier, we had passed the end of a similar mountain, and dimly off in the southwest stood two others. Each separately pierced the desert floor. The mountain slopes rose barren for two thousand feet, then had a broad belt of cedars, pines, and spruces. All day we had not passed a single tree, and the total number of stunted sages and bunches of grass was less than a score.

Stretches of the level lake bottom were yellow pavement of sun-baked, sun-cracked sediment. But most of it was covered with soda dust, sand borax, and salt.

On, Piute and I traveled, looking for a spring. By a barren water-worn rock we spent the night, without water in the dusty sediment of an ancient lake. Piute stood near me all night, watching my every move and depending on me to get him out of this predicament. Far off through the night we saw a light; probably the campfire of a prospector with burros high up on a mountainside, in the woods with grass and water. A red volcano in the eastern horizon and dawn was on the desert! Green fields and cool lakes lay just ahead—false promises of the mirage. By seven o'clock the alkaline dust was sizzling.

We aimed straight across the level desert for a spring twenty-eight miles distant. All day I saw what I had come to the desert to see—deserted shorelines and the dry bottom of a fossil lake.

Piute and I were tortured with thirst. A hot wind, fortunately behind us, showered us with powdery alkaline dust and filled the air with salty sand for hours. The storm ended, and I walked to save Piute, who had been staggering along. He stumbled, went down, and for minutes lay

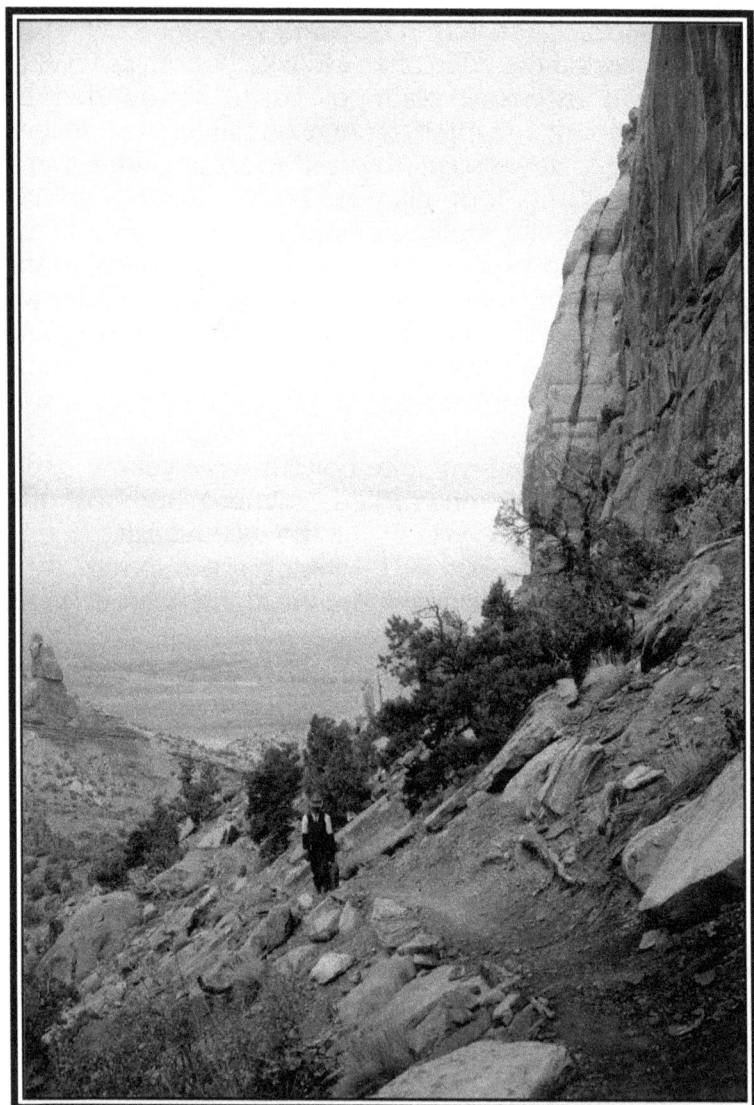

Looking out from the entrance of Colorado National
Monument. Photograph by George L. Beam.

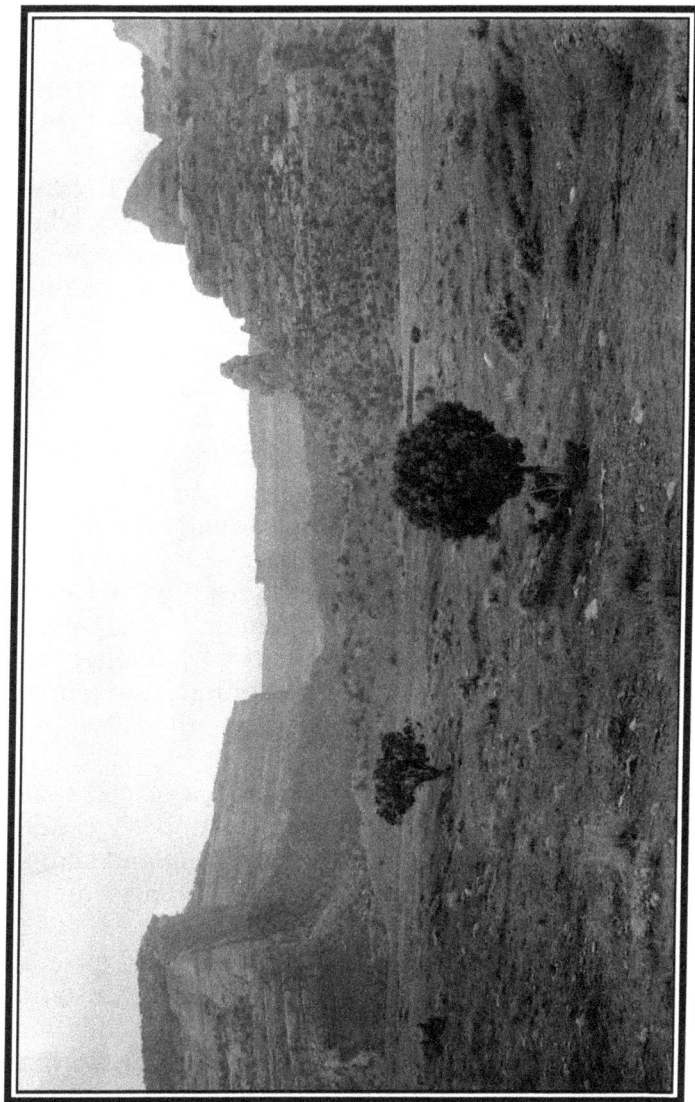

The Entrance to Colorado National Monument by George L. Beam.

groaning. After two desperate trials he rose and slowly followed me.

I was half blinded; the stinging alkaline dust burned my eyes. We were forcing ourselves along when Piute pricked up ears and turned gingerly off to the right. He headed straight for a mirage waterfall that seemed so close that one should hear its waters roar. But he was in home territory and brought up at a spring. The water was like Epsom salts, and we dared use but little.

Piute was restless, and long before morning I obeyed his urging, and we moved on. Evidently, he knew where there should be real water. He was hurrying forward when we met a prospector. Still maddened with heat and thirst, I called to him, "Where are we?"

"You are in the northwest corner of Nevada, with Oregon a few miles norther and California several miles to the west," he said.

"But how far to water?" I asked, while Piute was pulling and stepping about trying to go on.

"About a mile—your pony knows the way," was the answer.

After an afternoon and night at the spring, we started across a less barren desert for Malheur Lake, Oregon, where I was to see thousands of waterfowl. Mid-afternoon, Piute turned aside and stopped by a spring. I had planned to camp five miles farther along, but, rather than disappoint the pony, I camped here.

While stopping at this spring, a mirage placed a lake, teeming with waterfowl, just beyond camp. Geese and ducks were swimming in the water, feeding and sunning themselves along the shore. We traveled on to the real lake.

After two weeks with Piute and the desert, I gave him up at the lake and hurried away by speedier transport: a dry-as-dust vacation, but one decorated with mirages and many things for the imagination. As is common for those who know the desert for a week or longer, I went away planning to return.

During my next desert visit, light—the desert magician—showed another lake picture. Early one morning, a mirage lake appeared in the scene before my camp in western Utah. As I looked, a bighorn ram raised his head like a periscope through the silvery surface of the lake. The remainder of his body appeared to be submerged in the water. For a few seconds, his head also went out of sight, then reappeared.

There was a blur, and the next scene showed a ram, three lambs, and two ewes, all knee-deep in the shallow water of the lake. Shallow, short-lived lakes are common in the Great Basin. But how, a moment before, had the ram shown only his head, and where had been the others of the flock which now stood by him?

The ram walked forward a few steps, stopped, and turned his head. Others of the flock were starting to follow when the picture faded. After a few minutes, the lake vanished—but not the sheep. There on the desert, correct for distance and direction, stood the six sheep—a ram, three lambs, and two ewes—that had been in the mirage scene.

Evidently, the air was made up of layers of different density or of different humidity. The top of this obscuring layer must for a time have been just beneath the ram's head. Later, it dropped to knee level, or the sheep walked to slightly higher level. Here was a mirage stage-setting with real and undistorted figures in it.

From this camp, the following afternoon, a scene of different type was staged. It was intensely hot, and the sun seemed like molten metal in the hazy, coppery sky. Round me were level, seared, desert distances without a butte or a cloud.

A bit of seashore suddenly had a place in the hot dry landscape. A wave rolled easily in and flattened on the shore. Swell after swell, then breaker after breaker, rolled in upon the shore before me. Far out, I saw a heavy breaker coming in. It rose higher as it approached the shore, curled and broke almost at my feet. But there was no sound. It was uncanny. A transformation came so quickly that I could not follow. In apparently the same scene, without a breeze, a heavy fog bank came drifting in. The sun touched its edges to glass as it came on. For a moment, it obscured the sun. I said to myself, "This is exactly like real fog. If so, it will be moist and cool, wind or no wind." For a moment, there was gray obscurity; then, again, the soda-dusted sand dunes lay shimmering in the furnace air before me.

A mirage is the reflection of something; sometimes the mixed reflection of several things. It appears that an object or a landscape is lifted, perhaps by reflection, projected afar, and then set down in another place as a mirage. It may be of something near or of something miles off. It may be right side up or upside down. It may be photographically clear, or vague and cloudy, or of a confused mixture. This confusion may be due to several reflections mingling in the same picture, like several images being taken on the same negative.

Few mirage river pictures are a realistic and artistic success. The best one I have seen appeared to be a loaned section of the Platte River out on the plains—low banks and a thin flow that covered part of a wide sand channel. The remainder of the channel was of dunes, drift, and ripple-marked sand. The short squatty shadow of a long-armed cottonwood said high noon. I expected a cowboy or coyote to come into the scene, but none did.

Most of the striking and distinct mirages made for me their momentary place, their brief pause, in a western or southwestern desert. Museums are installing splendid natural history and other groups, and I hope that some time a grandly reproduced mirage will have a place of distinction in every large museum.

I was wanting a place to camp where there was wood. With canteens full, I was not concerned for water. In the dim distance to the right stood a cottonwood tree. As I did not want mirage firewood I looked closely before starting toward it. On the desert, one cannot always believe his senses. There are deceptive illusions. The mirage shows many ambiguous images. Desire often insists we are seeing the thing we want. There were two old cottonwoods, one behind the other. Both carried the usual mistletoe clusters in their tops.

But as I approached them, they leaped up and landed

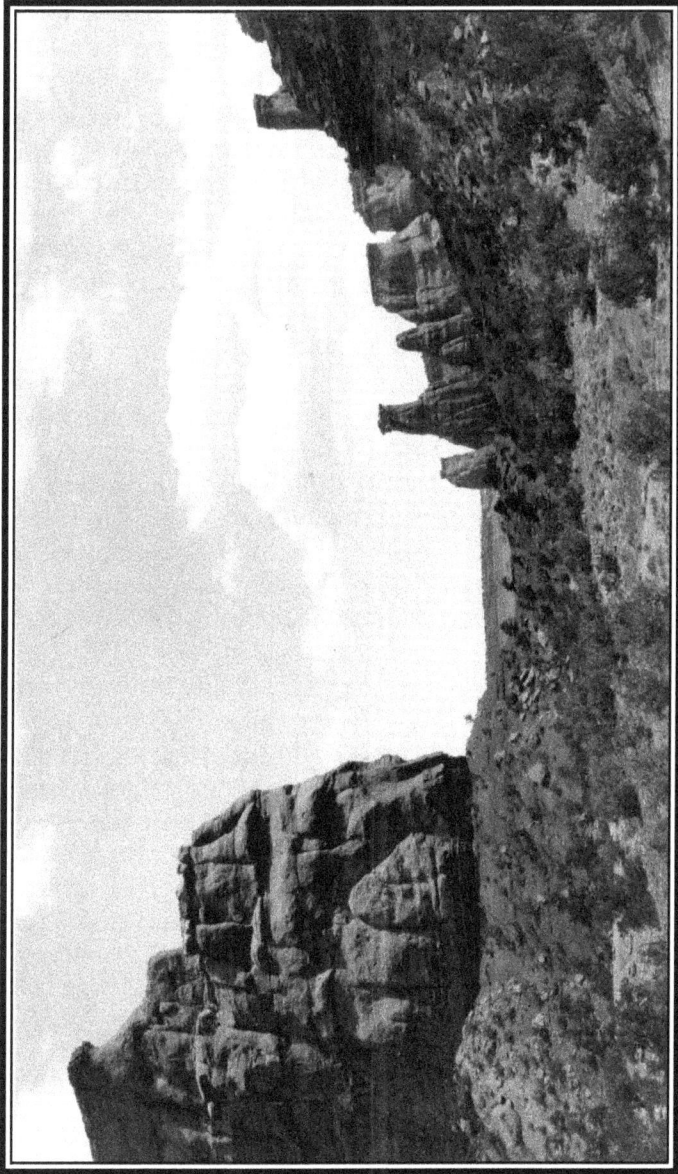

Entrance to Colorado National Monument.

on a steep nearby mountainside. Except that they shot upward, the performance would have passed for a landslide. But it was without sound. The trees sank slowly into the mountainside and I turned away to search elsewhere for firewood. Surely these trees must have been reflected from another horizon; anyway, they no longer stood in mine, and I commanded miles of distance.

Camels once, ages ago, inhabited American deserts. They are so frequently pictured with palms and deserts that it is not easy to dissociate them. If one be seen separately, the imagination either supplies the others or begins a search for them.

The camels are gone from the Southwest, but a few palms survive to give a touch of decoration and poetry to the desert's rim. In retreating from an adventure into Death Valley I looked eagerly toward the horizon where a cluster of palms with a proper topographic setting stood beside a pleasant spring. Near the seven palms were two camels. No one had heard of two camels in the desert, but prospectors are enterprising, and it seemed likely that a prospector might have discarded burros for these higher-geared transports.

I climbed over ridges of sand, and in due time arrived at the palms and the spring by the foothills. Two prospectors were in camp with two burros. But I had seen two camels: one with a pack on his back and the other with a pack on the earth by him. After a few drinks and an exchange of experiences with the prospectors, I edged out for a look at the camels. Not seeing them, I walked over to the spot where they were standing when I saw them across the desert. I could not find a track. I called to the prospectors, "Camels!"

None was at the spring, nor had there been.

All one morning I sat on a foothill in a California desert watching an antelope below. Behind me the mountains rose in steep, sharp-edged, naked rock walls. There were a few scattered shrubs in the short canyons. Beyond me, the desert was brown and level with widely spaced vege-tation. Distant sage zones appeared like dwarf orchards, and cactus groups like aggregations of weird, much-branched posts. Big ragged-edged spaces of sand in the sun shone dazzling in the purple distances. When the antelope moved out of sight, I left the water hole to go down and out upon the endlessly level desert. But, after a few steps, a desert water hole came into view, and I lingered to watch this institution. A shallow basin of water showed below a mere trickle of a spring. At two places a few steps from it, white bones were scattered. Close to the spring lay a carcass, apparently that of a burro. At this, three coyotes and twice as many buzzards were tearing and quarreling. Overhead, other buzzards were sailing. All this seemed intensely real, and it probably was. I did not go down to verify. Places in it, possibly sections of it, may have been modified or enchanted scenes superimposed over entire spaces.

Desert springs and water holes are far apart. All trails and air avenues end by them. To them come birds and antelope, rabbits and butterflies, skunks and snakes. The coyote comes to eat as well as to drink. Here animals and birds often rest for hours, and there they often play. This picturesque watering, gathering place, though often a life saver in the desert, is not always set in poetry. It shows records of grim feasts; it has a circle of skeletons.

The turning of my field glass upon a mirage often changed it into nothing—or formless light and shadow. Now and then, however, the glass gave me a correct focus on the object.

One day, steadying myself against a dead and thornless angle of a towering bent-armed cactus, I turned my glass upon a newly created and barren mirage mountain; but through the glass it was forested, and up a zigzag trail climbed a prospector with a pick upon his shoulder, leisurely following a burro with a pack. The burro stopped for a bite of something. The prospector stopped and used his pick. The burro lay down. The prospector rose up and looked at him. The burro started to roll. The prospector was hastening toward the burro when the mountain became barren and lifeless again.

The mirage can multiply and enlarge. One day, a small vague village of one street hung against a distant sky. This,

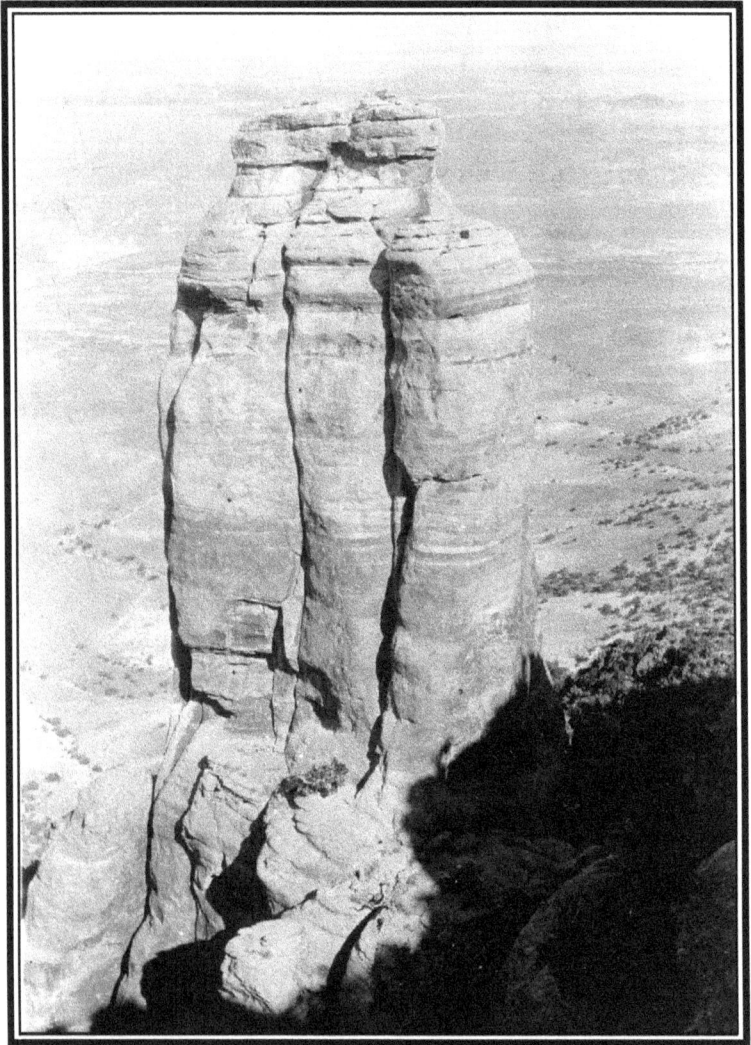

Pinnacle in Colorado National Monument.

if I be not mistaken, was created from twin, face to face, square-cut cliffs on a ridge near camp.

I left a ranch on the Carson River, Nevada, to walk across the desert mesa to another ranch. Late afternoon, I appeared to be arriving, for below me lay a stream, a fenced yard, a house, and a scattering of trees. This answered the description of the place sought.

A dark storm cloud with streams of silver was approaching the sunny ranch. Lightning, golden crooked rivers of it, flashed against the dark background of cloud. Everything grew darker with the advance of the storm. Lightning flashed splendidly down and appeared to strike a cliff on the mesa behind the house. I stopped and listened, but thunder did not crash or roll. Slowly, the dry mesa with its stunted and scattered sage returned, and I went on looking for the ranch.

Reflection and refraction staged a scene before my Utah camp that caused me to do some reflecting. I was on the frayed outstretched margin of the desert: foothills half covered with scanty grass; acres of prickly pear; stunted and thinly planted cedars and pines; sandhills; gorges with southerly facing walls bare and brown, and northerly ones carrying a green tone of vegetation. I had just examined a nearby cliff with my glass, and on lowering the glass, a stout grizzly bear came walking along where a second before no bear was present. Along the bottom of the cliff, which was a block in length, he walked toward me.

His movement, size, stride, and color were correct. He was a grizzly.

"A mirage can make a desert blossom, build mushroom cities, and exhibit prehistoric life," I said to myself. "But it cannot, when I am watching, slip a real grizzly into the scene right under my nose."

Theoretically, the mirage cannot do this, but there was a real grizzly, and almost close enough for an introduction.

"If that is a real grizzly," I said to myself, "he will run if I yell."

He was close enough for me to see a scar over his left eye. Suddenly, he stopped and rose up; he sniffed as though scenting me. He should have seen me, for I saw

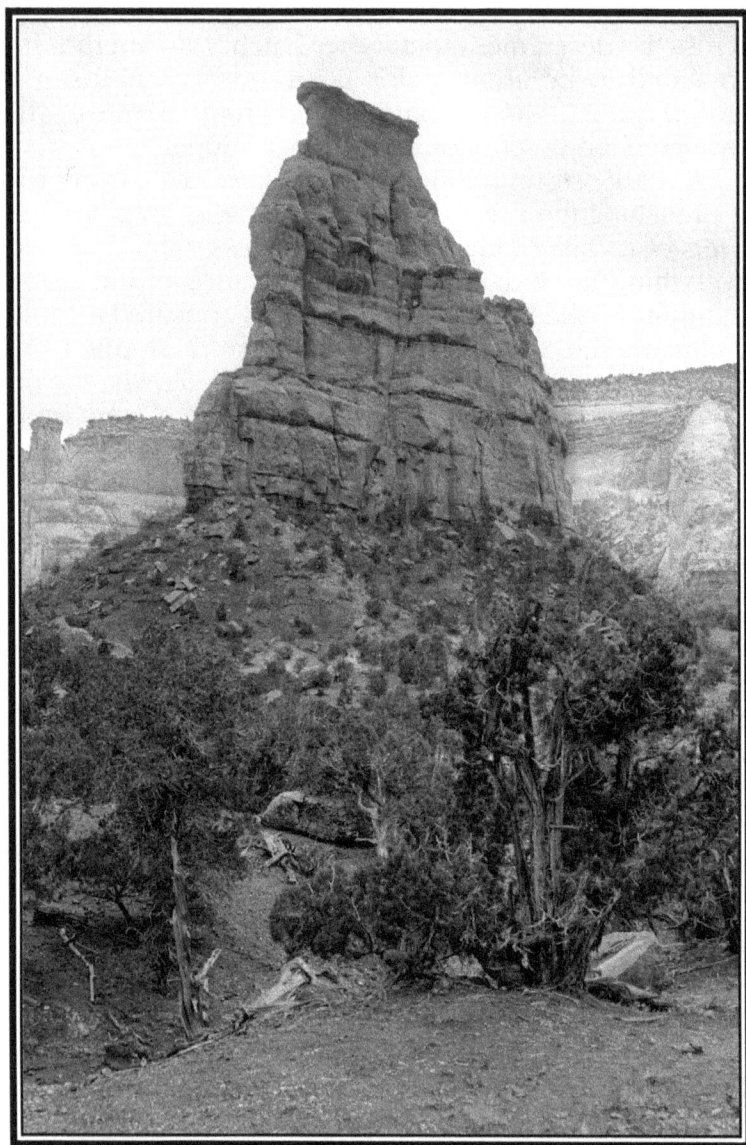

Independence Monument, Colorado National Monument, scaled by John Otto. Photograph by George L. Beam, circa 1915.

Another view of Independence Monument.
Photograph by George L. Beam, circa 1915.

him. I yelled. He nearly fell over backward. He retreated with a rushing gallop. The third or fourth jump the mirage fell to nothing. There was a cliff, sphinxlike, but not a bear in sight. I had not seen a bear. I walked round the cliff. Behind the farther end, I found fresh grizzly tracks in the sand. The mirage had bent the light rays round the corner as it were. I don't know just how it was done. I saw the bear.

A mirage is occasionally seen over lakes, plains, and the sea, but the desert it a prolific producer of mirages. On the desert, the mirage is weird, splendid and uncanny. But it is also peaceful and artistic. I have wished that more artists might see the mirages of the desert.

Apparently, desert air is the environment of the mirage. It is a magician and master conjurer. It produces air castles, green fields, bits of paradise. Earthly horizons and landscapes in its legerdemain are mixed with mystery and color. The mirage is an artist—it creates new landscapes from old.

Sight-seeing by wireless likely is near us. Ere long, an Edison may seize and transport these mirages by science and show them as they are before assembled audiences.

More About Mirages

All one winter day that I snowshoed across a high, level, white plateau, I was entertained by mirages which appeared before me, to the right of me and to the left of me. This mountain plateau seemed much like the treeless plains that went away white and level to every horizon. It was more than twelve thousand feet above the seashore and nearly two thousand feet above the timberline. Behind me, my snowshoe tracks led back into measureless snowy distances. Ahead, over the smooth white swells of the plateau, not a track or a shadow could be seen.

Suddenly, off on my left, two perfect spruce trees towered splendidly upon a snowy skyline perhaps three miles distant. The uppermost trees at timberline, far below, were small and ill-shapen, and not so tall as one's head. The conditions on this arctic-zoned plateau told me that even tiny dwarf trees could hardly live here. But, looking through my glass, the two were the best impersonators of Engelmann spruces that I had ever seen. I headed straight for them.

After advancing about a block, there was a shimmer of light and the two spruces melted. Where were they from? I stood waiting, thinking that they might flash back. But they did not. Before me, two slender grass stalks stood above the snow. I circled back to where I had first seen the spruces. They were in view again, but this time they were upon a snowy rim of a canyon—the magnified overlapping snowshoe tracks that I had made by the grass stalks.

A little later, when a real arctic ptarmigan, pure white, stood upon the snow within fifty feet, I stood and watched it for some time, unable to judge if it were real or a magnified and reflected bit of shaped snow. Nor could I figure the distance; twenty feet, perhaps. Then, as I looked, it certainly must be a quarter of a mile, but finally I thought I might touch it by bending forward. The fifth step I took toward it, it ducked into its snow burrow; it was seen at seven steps.

An hour later, a ridge purplish black with forest stood out in the air on my left. Beneath and beyond, the real mountains could be seen in place. The mirage did not rest on anything or have any attachments; it was detached and afloat in the air, but absolutely becalmed. Presently, it shimmered in the heat waves and then melted into the viewless air.

Half a mile ahead, the level plateau ended in a narrow peninsula-like projection that stood a thousand feet or more above its surroundings. On the right of this point, a towering snowy peak suddenly flashed into place, attached to the plateau without a flaw. For a moment, it reeled and flickered in peculiar light, then stood still—a snowy summit old in story. Suddenly, a snowy peak had a place on the opposite side of the plateau. I could not tell where it came from, but my ears listened for the sounds which should have been loosened with its creation. Serene and beautiful, it stood with its cliffs casting rough-edged and bluish shadows over its drifted snows. With a lurch, it sank, and the peak opposite rose. Then, with a tilt, the first one rose again. Up and down, they rose and sank, teetering as though upon an invisible support laid across the peninsula. Sometimes they balanced or swung back and forth slightly as they seesawed. Though they had no visible means of support, they alternated perfectly. Having exhibited their specialty, they vanished in a shimmering screen of air waves.

As I stood on the edge of the plateau trying to account for them, where they had been imported or perfected from, there appeared, off in the distance, a heavy yet graceful long ice bridge across a mountain canyon. The heavy pillars were ice; the arches, the banisters, and the roadway were all of ice. Beneath the wide middle arch, floating ice cakes crowded the flowing river.

Mirages are common on deserts, on high dry mountains, in arctic regions, and are sometimes seen on wide plains and on the plains of the sea. Deserts have the most frequent displays, and then, too, mirages are more varied on the desert; they often feature lakes, streams, and the green luxuriance of moist mild climes. Deserts also often exhibit mixed mirages—several scenes focused at the same instant on the same space.

White mountaintops and the icy arctic are more given to the magnifying distorting of nearby objets or the placing of new mountain ranges above the horizon's far-off rim. In the Rocky Mountains, there is an area equal to that of New Jersey, Maryland, and Delaware that is entirely above the timberline and more than two miles above sea level. Here are extensive moorlands, arctic tundras, snowfields with arctic flowers, peaks, glaciers, lakes, and canyons. In this region, in winter, is often exhibited magic and spectacular scenery—the mirage. This mirage sometimes confuse explorers of the heights by adding a mountain range to local topography or a scenic screen causing local scenery to appear as a part of the cloud scenery of the sky.

During a snowshoe trip across the Colorado Continental Divide I had traveled two hours since leaving timberline on the eastern slope without arriving at the summit. It was late in the afternoon and clouds were beginning to envelop the surroundings. After passing a cliff where I expected to look down on the slopes below, I was confronted with miles of snowy plateau which in the distance went up in a snowy peak. This could not be passed before dark, so, making a hurried sketch map of it, I turned back to timber to camp, so as to have a full day for crossing the summit.

The next morning, on arriving at the place where I had mapped the plateau and distant peak, I found these missing, and within a stone's throw was a canyon which dropped away rapidly down to timber on the Pacific slope. I had mapped a mirage.

Was this mirage mountain a bit of broken lowland magnified and uplifted, or was it a distant bit of mountain that had been transferred—placed by reflection? Anyway, it proved a most deceptive piece of background scenery.

A mirage is a moving picture, a reflection of the real—by wireless. Like a fragile package, it sometimes is thrown out of line and distorted in transit. It may be magnified, multiplied, minimized, or mixed and arrive with other somethings, but, as started, it was one hundred percent real. Often it joins the local scenery as though a part of it.

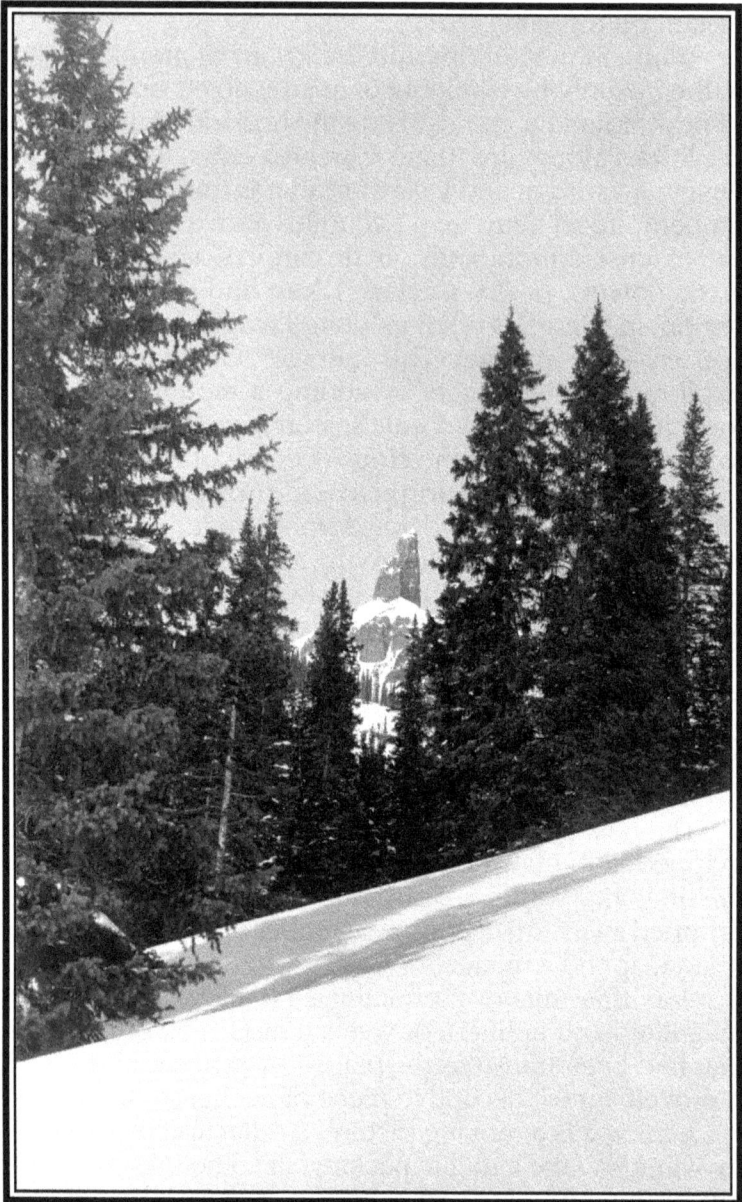

Lizard Head from Rio Grande Southern Railroad Track,
Feb 2, 1911. Photograph by George L. Beam.

Lizard Head from Rio Grande Southern Railroad Track, Feb 2, 1911.
Photograph by George L. Beam.

In all the places where the mirage performs with artistic legerdemain, the effects are often so realistic that the apparent scenes seem a part of the local topography. Indians, prospectors, sailors, cowboys in new scenes have mentally recorded parts of the mirage as parts of the landmarks of the region. Surveyors have mapped them, mountains climbers have set off to scale Mirage Peak. Under the Northern Lights the mirage has deceived the greatest explorers from many nations; islands leagues long and mountains high have passed skeptical scientists and censors and taken a brown place on the map.

For nearly a century, several Northern explorers had given reasons for the existence of an extensive land area to the northwest of Grant Land. In 1906, Peary's Eskimos reported seeing it; and in June of that year, Peary satisfied himself on this, named it Crocker Land, put it down on the map and wrote of it in "Nearest the Pole." He placed it about one hundred and twenty miles northwest of Grant Land.

He heard of a new land far to the north, and rest was not his until he saw it. In July, 1913, the steamer with the MacMillan explorers aboard moved northward, and their first object was "To reach, map the coast line, and explore Crocker Land, the mountaintops of which were seen across the Polar Sea by Rear Admiral Peary in 1906."

After traveling one hundred and fifty miles northwest to where he should have been thirty miles inland, MacMillan asked himself: "Could Peary, with all his experience, have been mistaken?" But, "April twenty-first was a beautiful day; all the mist was gone and the clear blue of the sky extended down to the very horizon. We ran to the top of the highest mound. There could be no doubt about it. Great heavens, what a land—hills, valleys, snow-capped peaks extending through one hundred and twenty degrees of horizon! The landscape gradually changed its appearance; finally at night it disappeared altogether." The following day was clear and "we scanned every foot of the horizon—not a thing in sight, not even our almost-constant traveling companion—the mirage. We were convinced that we were in pursuit of a will-o'-the-wisp."

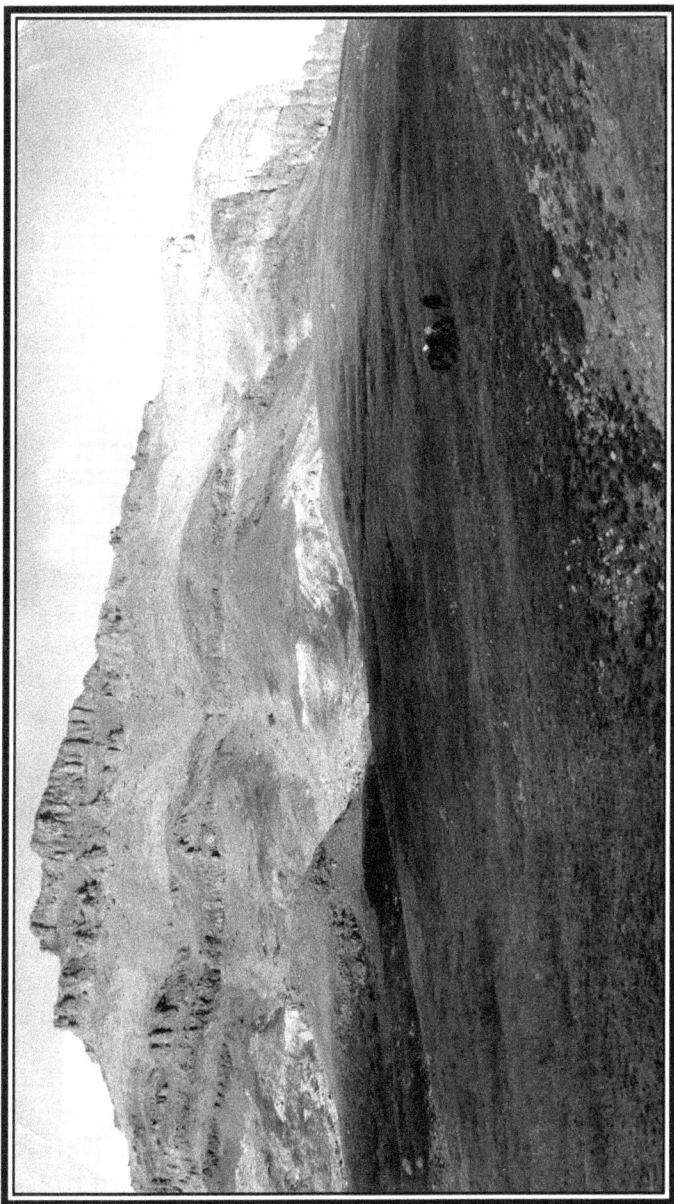

"A Mountaintop Moorland, 13,000 Feet above the Seven Seas."
Photograph by George L. Beam.

There are heat waves on mountaintops and in the icy north. These may shimmer over icefields with the thermometer showing forty below. Under these conditions, the mirage is likely to be the master shifter of scenes; the distant icefields are lifted into mountainous islands and rounded snowfields become hills and valleys. But sometimes they are transformed into fields of towering cacti or forests of giant mushrooms with rounded stems of varying lengths.

The wireless movie often brings far-off scenes into the arctic. During the Franco-Prussian War a number of scattered and independent observers in northern Sweden and Norway saw mirage armies marching through the air, equipped like the real ones that were fighting a few hundred miles to the south.

A range war in central Wyoming was nearing the end, and one combination of cattlemen was making a drive on the opposition, which was holding the fort in two ranch houses. The attacking parties were pressing matters vigorously when the besieged saw coming reenforcements already overdue and sorely needed. At two points on the horizon, oncoming parties were identified as friends.

Hoping to surprise the besiegers, whose position did not enable them to see these reenforcements, the besieged sent messengers to the coming reenforcements and signaled them where to strike. Then, out from the ranch house they poured. This surprised the other side, but it did not stampede it. So rapidly did it throw lead that the besieged, though not captured, were driven off and the ranches taken.

But where were the reenforcements? They were coming, but the mirage had served the enemy by causing the two bands of cowboys to appear twenty or more miles nearer than they were.

One snowy day out in the level white plains of Wyoming, I saw smoke on the horizon. Going to the highest point, this appeared to be rising from the sod house of a settler. A wagon stood near, and behind the house were two horses on the sunny side of a dilapidated straw barn. Apparently, this sod house was two miles

The San Juan Mountains: Uncompahgre Peak at right; Mount Wetterhorn, center; Mount Coxcomb at left. Photograph from summit of Engineer Mountain. Photograph by George L. Beam.

distant, but I knew that the nearest one of this type was in western Nebraska, perhaps one hundred miles away.

While fossil hunting to the north of Red Desert in Wyoming, I noticed, one morning, a dust in the sky as though a herd of cattle was below the horizon. Presently, I saw the trail boss lolling in his saddle as his black horse stood in sleepy pose. Then the herd came into view—part of it—the upper part—for of the several hundred pairs of horns showing, most of the owners were slit horizontally through the body and were without any legs; yet they were sliding forward. A moment later, heads and upper parts of their bodies vanished, and their legs, with a narrow slice of the lower body, went legging across the dusty plain. All the time, the cowboy trail boss lolled in his saddle in view, and I could not understand why he did not sit up in astonishment at the cutting of his herd to pieces. But the fact was, the entire herd was behind the horizon about twelve miles distant.

Mirages give almost a continuous performance somewhere on the desert. They are of infinite variety, and—like camera pictures—of every degree of excellence, including the showing of several on a single negative and those in focus and those not.

A vague or confused mirage brings interesting comments, and often reveals something in the mind of the onlooker. Once, while I was talking with a number of covered-wagon emigrants on the desert road near Ogden, a dim mirage village was shown. A minister in the party at once saw a chicken yard; a mother looking for a lost son wondered if he could be in this place; while the figure of a gigantic skunk was pointed out by a gentleman of the party who, in the dark night before, had stumbled over a black-and-white chemical extinguisher.

A geologist long remained in camp in the Great Basin. Here are piled broken geologic ages which are readily seen—there being no clothes, not even concealing drapery of vegetation.

One evening, the cowboy of his camp rushed in with: "Get on to that dinosaur group! The whole Cretaceous period is on exhibition—the age of reptiles is giving its last show."

There it was, and even the cold and exacting geologist confessed it a good prehistoric group of the reptilian horizon.

During an interview with Clarence King he said to me: "My first sunset was in the barren desert of central Nevada. As evening came on, those naked rhyolite rocks shone in more glorious colors than any autumn colorings I have ever seen. The peaks stood out in sharpest lines of silver silhouette, then came broken and colored clouds on the horizon. As the sunlight failed, one of the cloud mountains became realistic sky scenery—a volcano with throat of red and black that showed a broad lava flow of molten rubies against the sky of night."

Many of the desert mirages have a distinct artistic quality. Many unimportant details are lost and striking features are intensified by exaggeration or position. Many well-known artists have commended the desert because of its artistic and striking suggestiveness, and one well-known artist who has given many glimpses of desert landscapes declares that his best picture is about ninety percent a copy of a transient but telling mirage.

The mirage frequently uplifts islands and vessels from below the horizon, placing them against the sky or a background of clouds. There seems to be less distortion of the sea pictures than of those on land, and seldom are the pictures piled on several deep and confusingly, as often is the case in desert mirages. But sometimes in a sea picture an upraised island is shown in duplicate, one reflection resting upon the other. Generally, the image shown is not moved to one side, but just uplifted above the horizon's rim. That these mirages often are directly above the real, I have proved with islands in the Pacific by taking a compass course and sailing directly to the real island.

Deserts have a monopoly of the jazz revels of dust devils which dance in the heated air. But they have one thousand and one other displays, and often exhibit scenes from side lines.

"Beef on the hoof!" called the cook, and bacon-fed fossil hunters rushed from dusty tents. It was several weeks since fresh beef or any fresh food had been served at the camp among the desert sand dunes. The teamster of one load was misled by a mirage and lost, and the second load was overdue.

A whole herd of cattle appeared to be rapidly trailing and grazing across a distant grassy hill. I had just arrived in camp and there was no such hill within two days' journey that I could recall. It was a mirage masterpiece. But nearly everyone in camp declared the scene must be a real one, perfectly reflected from somewhere. The cowboy —horse wrangler—had been watching without a word.

"How about this?" said the artist, turning to him.

"A good picture," he replied. "But the critters are too long for their height and look like the work of an artist who was not strong on anatomy."

The artist and everyone laughed, but all kept eyes on the animated hill. Presently, one of these live "critters" caught another in its mouth and walked off with it like a dog with a big stick. While we were staring and commenting, another one of these movie cattle fell backward off a cliff, struck, bounced, and then picked itself up and hurried after the others as though nothing had happened.

Then the cowboy burst out with: "I guess she is projecting ant hills today!"

Several times, while I was in this camp, we had semi-mirage effects. One evening, the cowboy was bringing in strayed horses. Halfway up on the sky, we saw him and two unwilling horses reflected. One horse broke back; after him the wrangler rode, swinging his hat. The other horse paused to watch the result. Then all advanced towards us, but turned to go up an arroyo. All this time, they were projected into view, but in reality were out of sight behind sand dunes. As they passed over the top and into direct range of our eyes the transition from the reflected to the real was but a mere break barely detected.

The cowboy on horseback had paused one evening on a nearby skyline sand dune to watch a coyote. He threw his leg across his horse's neck, and there he rested in cowboy attitude. In the twilight we saw him, but in a mirage scene. There were the plains about him; a cowboy camp wagon in the distance, with the herd beginning to lie down; then, as the wrangler rode away—around the cattle a night herder. But on return to camp the cowboy declared that he had seen none of these.

A few years ago a well-known minister held services among the picturesque granite rocks on Sherman Hill, a few miles west of Cheyenne. The day selected was an ideal one in June. Several hundred people, a number coming more than one hundred miles, were on the ground waiting for the sermon. It was already time for the service to begin, but the minister, with car trouble a few miles away, had not arrived.

But he was on the way. A number of people climbed to the top of the highest crag, hoping to see his car rushing in. They saw it, and it was giving a spectacular performance. Off in the northeast was a broad cumulus cloud, and over the surface of this his car seemed to be dashing. He was doing things on high. He was registering with sky pilots. His car leaped several canyons in the clouds, and at the edge of the cloud where this blended with the earth, the car sailed over it like an airplane coming into a landing. On arriving, the minister admitted speeding and having seen a minor mirage, but denied knowledge of a movie performance.

The mirage often attaches itself to something in a stage effect. One evening, near camp, there suddenly appeared a new white tent; by it two men moved about near a campfire. The camp was by the edge of a lake. All of us knew that there was no lake near, but the whole scene was so effective that all went out for a look in the morning to see if it had lingered. The men and the tent were there but the lake had vanished.

One of the most realistic mirages that I recall I saw on the desert to the southwest of Great Salt Lake. A mountain with a snowdrift cut into the sky; a forested slope and a canyon with a concealed river reached out to the fore-ground, where a splendid waterfall leaped in white glory. By it, standing close to us, was a scattering of spruces in a meadow opening. The edges of the picture were ragged and irregular, but shaded off gradually into the desert or the dull background so that the effect was startling and magnificent in the extreme.

There were a number of people in camp at the time, and all saw the same picture. There was nothing indistinct, nothing ambiguous, that one individual resolved into one figure and another person into something else. A dog of his own accord looked at it with every show of surprise and uneasiness, and two horses gave one look and started eagerly off in that direction. Explain it by alchemy, legerdemain, or hypnotism if you will, but this was a scene so real that, though it came on suddenly, it deceived both man and beast. The scene probably is somewhere in Utah, but it could not be mapped as visible from the place where we saw it.

Trailing Utah's Shorelines

I assumed that everyone in Utah knew the wonderful story of old Lake Bonneville. But the hotel clerk to whom I mentioned the historical ancestor of Great Salt Lake looked up from the morning paper with a fossil stare—nothing more. A First Settlers Club referred me to the State Historical Society, but this organization, dealing in recent human affairs, was not even interested in prehistorical geology. And the secretary of a Chamber of Commerce suggested, kindly but firmly, if I had in mind the publication of a story concerning a wholly imaginary lake, that I locate it in some other region, as they did not care for that kind of advertising.

I had failed to impress anyone with my enthusiasm for trailing ancient shorelines when I at last encountered an old prospector outside the city limits.

"Yes," he said reflectively, "old Bonneville was a thousand feet deep and covered half of Utah. Its waves wore an enduring shoreline upon the slopes of the Oquirrh Range; and, in fact, all the Great Basin shows evidence of the old lake level, high and dry on the mountainsides. Lake Bonneville was fresh water. Its sediments and its abandoned shores are strewn with freshwater fossils. This immense, dried-out lake basin, in which half a million people now have their homes, schools, farms, mines, and factories, in the largest fossil known."

After a few more illuminating remarks concerning this dead lake and the records which it had left in shorelines, sediments, and evidences of long submergence, the prospector turned away and I started off for the Oquirrh Range.

As I skirted Great Salt Lake, which occupies a small area of the old lake bottom, the shore seemed strangely like that of the sea. Gulls were flying over and waves were forming and thundering along the beach. Deltas had formed where streams poured in, and on beaches and headlands winds, like happy children, were building and shifting their piles of sand. The breezes had a goodly tang of salt, as though sea rivers were flowing into the lake; but none flowed out; its waters left only in clouds.

The old shorelines showed boldly in the distance, like highways along the mountain walls and around mountain peaks that had stood as islands in the ancient island sea. In places a dozen shorelines, one above the other, formed gigantic terraces. Each marked the level at which the slow-changing surface of the lake remained sufficiently long to wear an enduring record. The many shorelines tell of a fluctuating existence, of ups and downs, while old Bonneville had a place in the sun.

I climbed up for a closer view and, near sundown, paused in a gulch on an old shoreline wide enough for a four-track Lincoln Highway. Here was the high-water mark of Lake Bonneville, one thousand feet above the surface of Great Salt Lake. At its height, Lake Bonneville comprised about nineteen thousand square miles—the area of Lake Michigan—and covered more than half of Utah and a small area in Nevada and Idaho. It was about three hundred and fifty miles long and one hundred and forty-five wide, with a three-thousand-mile shoreline stretching across desert plains and mountains and extending far back into mountain canyons for slender bays. Mountaintops large and small were its rocky islands.

As darkness came on, I made camp on this ancient beach line, alone and happy—my campfire and I, beneath desert stars, by a fossil lake on which the waves had not rolled for at least twenty thousand years, possibly not for twice as long. For years I had been hoping to explore this long-deserted ruin. As I thought of trailing the old beach line, my imagination was stirred. While the waves of this vast lake were washing and wearing its boundary, there were elephants, camels, horses, and other mammals feeding along its beaches or trampling the headlands above its shores. Possibly, primitive man was also in the scene. In the sediments of nearby fossil Lake Lahontan a primitive spear head has been found beneath twenty feet of sediment. For thousands of years, the old shoreline has endured wind, frost, and rain; running water and landslides

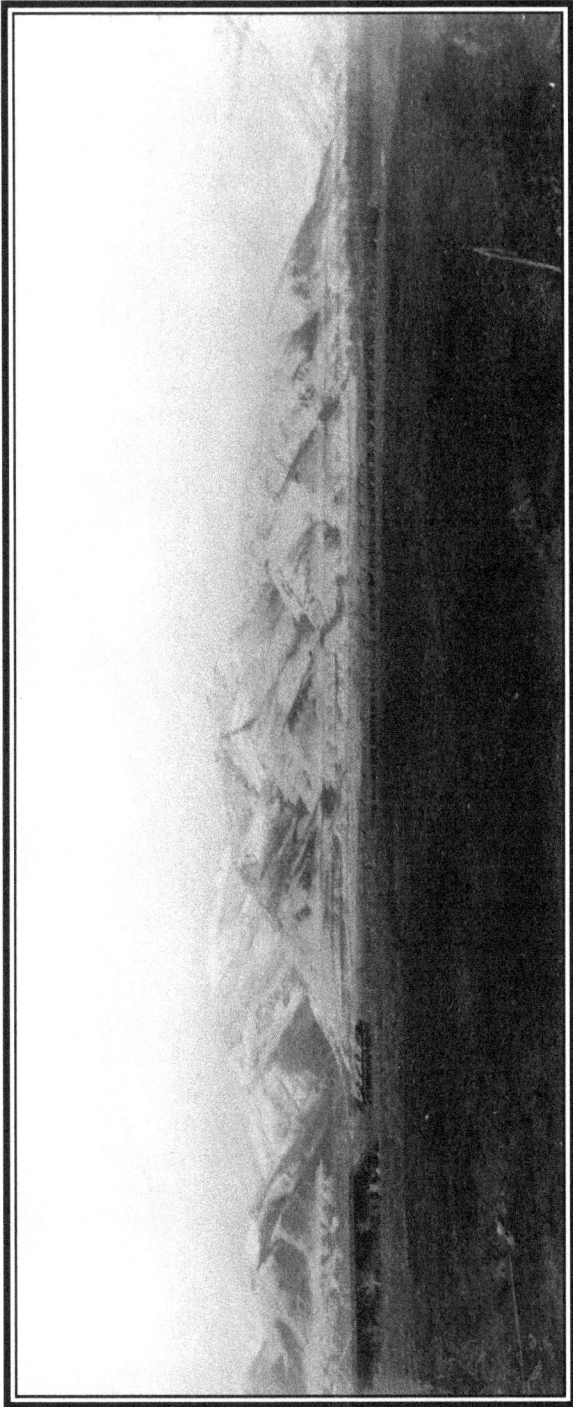

Lake Bonneville water line markings along Wasatch Range southeast of Salt Lake City.
Photograph by Frederick J. Polk, Ph.D.

have also done their work. Today, modern man fills and stirs the scene of this geological story. Miles of radiating railroads with their bands of steel are firmly bedded in the bottom of this deserted and deserty lake bed.

The next morning the shoreline highway brought me out on a long, outthrusting ridge that had been a peninsula in the old lake. As the shoreline folded around it, I looked down on the shallow shore of Great Salt Lake miles away and a thousand feet below.

Immediately below the Bonneville shoreline, on which I stood, the mountainside was terraced with five shorelines in close succession. These, so the wave sediment showed, were made by five pauses of the rising surface of the lake. When high water reached the Bonneville shoreline, the lake found an outlet north at Red Rock Pass, Idaho. Through this outlet, the waters eagerly rushed, rapidly deepening the outlet. In a short time, it was cut three hundred and seventy-five feet down to the Provo shore-line—the boldest of them all. At the Provo level, the lake surface appears to have lingered many thousand years. Finally, it sank below the outlet, lingered, made other shorelines, and became salty.

I climbed down to reach the Provo shoreline and followed it around the mountain all day long. Long segments of it were missing; these had been washed off by running water or had slipped away with landslides. Stretches were covered with landslide debris or sediment that had washed and rolled from above. Wind erosion and windblown sand in places had cut it to pieces or had buried it. But, though broken, miles of it were so well preserved it seemed almost incredible that the waves had not splashed it for ages.

Back in a canyon I found the remnants of a delta. Here a small stream flowing into the lake had dropped its burden of sediment—an enormous dump. After the lake lowered, the stream had washed a part of the delta away. These deltas, with upper surface roughly level with the old lake level, had filled in and built outward at the stream's mouth just as deltas are formed today.

As I walked along, noticing gulleys, eroded deltas, and

"The lurid and many-colored clouds over the desert horizon made a sunset worth crossing the continent to see." Great Salt Lake, Utah. Photograph courtesy of Utah Photo Materials Co.

debris piles left by streams and landslides, I became conscious that the shoreline was tilted, and that the water had run over it in the same direction in which I was traveling.

The arrangement of accumulated sediment on the surface of the shoreline showed that at a low point streams from opposite directions—from the east and west—had met, then overflowed in this break.

This warping puzzled me and I accounted for it with local landslip and local cave-in. Made by the horizontal surface of a lake, these old shorelines were originally dead level. The Wasatch Mountains are still rising. Many faultings and warpings of the crust of recent date are seen. A water-pipe main was recently broken near where the pipe crossed a fault in the underlying crust.

One night, I camped in a sheltering niche of the Provo shoreline. The lurid and many-colored clouds over the desert horizon made sunset worth crossing the continent to see, but gorgeous horizons are common in the Salt Lake Desert.

Great Salt Lake now is about six hundred and twenty-five feet below the Provo shoreline. It has an average depth of thirteen feet. Bonneville was a thousand feet deeper. Salt Lake, like Bonneville, has made many shorelines and the records of its rise and fall have been kept since 1850; twice—in 1903 and 1905—its surface dropped a few feet below the shoreline of 1850; twice—in 1858 and 1877—it rose ten feet above this level. Its surface now is several feet above the level of 1850.

The next day, I left the Provo and went out into the desert plain. Here the mirage put on an exhibition of old shorelines for my especial benefit. Evidently segments of real shorelines had been reflected and projected. Lines from the Promontory Mountains, the Oquirrh Range and Fremont Island were shown, mingled, magnified, and wrecked with greatly exaggerated tilts and depressions.

Unexpectedly, while crossing the old lake bed, I came upon lava that had poured red hot into the lake. Volcanic action has now and then been recorded in the Basin since its creation. Volcanic ash shows in many layers of the lake's sediments. The spectacular part that had been played by these old craters was as exciting as any of the lake's records. One of the most telling volcanic exhibitions was near Deseret, Utah. One mound-like crater stood out in the desert plain belted with two shorelines. Lava was buried beneath the oldest Bonneville sediments; an old shoreline, evidently made by the first Bonneville existence, had been partly covered by lava of later date. Then, too, lava had flowed out into the water during the second stage of the lake and on the dry desert floor.

When miles out in the desert plain, I turned for another look at this fossilized volcano. Boiling over the top was a black, convulsed, and fire-tinged mass like a rolling flood of lava; while, still staring, a heavy cloud like smoke and ashes settled before me. Was I seeing the beginning of another volcanic outburst? I was ready to believe it when the scene shifted and I added another speciality to the master producer of movies—the mirage.

A year later, there was another mirage demonstration. This was by the Promontory Mountains to the north of Great Salt Lake. I had climbed one of the many mountains which had once stood a rocky island in the ancient lake. Its base was deeply buried beneath sediments. Across the far-reaching sea of sand rose many island-like, barren mountains in the barren Basin.

The entire scene before me suddenly became a mirage. With a rush, the old lake was restored. Most of Utah vanished beneath clear water. The surface of this mirage lake appeared to be a few hundred feet beneath me. I moved to another point and had a glimpse down into camp and of a young geology student who had been with me for a few days. He had been examining a delta near camp for gold, and now loafed about as though waiting for a train.

I called to him to look at the mirage. He looked, but plainly saw nothing unusual from where he stood. He started up, and when about one hundred feet below me called: "Did you see the reflection?"

When I looked again, the mirage lake was still perfect. Not a wave moved along the rocky shores of the peaks that pierced it.

However, not seeing any reflection, I hurried down to him and there reflected was the Wasatch mountain range and the white clouds in the sky. Often I had seen a mirage simulate lake or a section of the sea. Here was double deception, perfectly reflecting mountain and horizon, the clouds and the sky.

Lake Bonneville may be said to have been twice upon the earth. During the Ice Age, the flood of water which followed the melting brought Bonneville into existence for its first stage. It appears to have had a place in the sun for several thousand years and then to have dried up altogether. After a long period with the basin dry, it was again filled during the melting of the ice which followed the last glacial advance.

Though there were many down-reaching, out-reaching glaciers, at one place only do these giant ice tongues appear to have reached the shore of the lake. This is in Little Cottonwood Canyon. Here the shoreline deposits and moraines mingle; from the steep, sloping Wasatch, the Ice King probably launched many an iceberg on this island Bonneville sea.

Bonneville's first existence appears to have been long. The water rose slowly, with many fluctuations, to the depth of more than nine hundred feet. Though within ninety feet of a possible outlet, it slowly shrank and dried out altogether without sending its waters to the sea.

The sediments deposited during the first existence of the lake were deeply eroded by running water, and in places quantities of debris were washed and blown upon them, before the sediments of the second existence were deposited.

Its second existence, shorter than the first, was more of the nature of flooding. Its waters rose speedily to the depth of several hundred feet, paused briefly, rose higher, paused again, and finally rose to the depth of more than a thousand feet. This was the Bonneville level and this allowed the waters to flow north out through Red Rock Pass.

Eagerly the long-imprisoned waters made their escape though this high mountain pass, thence through the Snake

Bryce Canyon.

and Columbia rivers into the sea. This is the first time that water had flowed out of the Great Basin, perhaps after the existence of a million years.

That Bonneville was a fresh-water lake is shown by the presence of several species of fresh-water fossils found at varying depths in the sediments of the lake bottom. During the existence of Bonneville, Lake Lahontan was in existence in Nevada. This Nevada lake had long existence and no outlet, but during its long, deep stage it too was a fresh-water lake.

The Wasatch Mountains were composed of sedimentary rocks—rocks formed in the salt waters of the sea and which carried a percentage of salt. The erosion and weathering of the mountains gave a percentage of salt in solution to the streams that flowed from them into the basin. This salt must have accumulated as the repeated lakes dried out in the basin before the coming of Bonneville. With the basin lined with salt deposits it would appear that Bonneville should have been intensely salty from the beginning. But before Bonneville came, the salt that lay deep in its basin had been buried beyond the reach of its water.

During explorations, I came upon a number of small salt or soda deposits in the beds of extinct lakes. A few of these deposits were deeply though not completely buried. One deposit of salt was more than one half covered beneath a twenty-one-foot layer of drifted sand; another deposit was nine tenths covered beneath a twelve-foot layer which was forming from the sediments abundantly washed in during floods from cloudbursts. The natural cements in this sediment would in time form over the salt a rock strata that no surface water could penetrate.

Deeply buried in sediments are the bases of many mountains which pierce the surface of the old basin. With buttresses, lowlands, and approaches buried, they stand like mountaintop islands in the sea, and in Great Salt Lake.

A few years ago, the Southern Pacific Railroad drilled many deep wells seeking water in the desert Bonneville basin miles to the west of Salt Lake. The drilling penetrated lake sediments for from several hundred to a thousand feet.

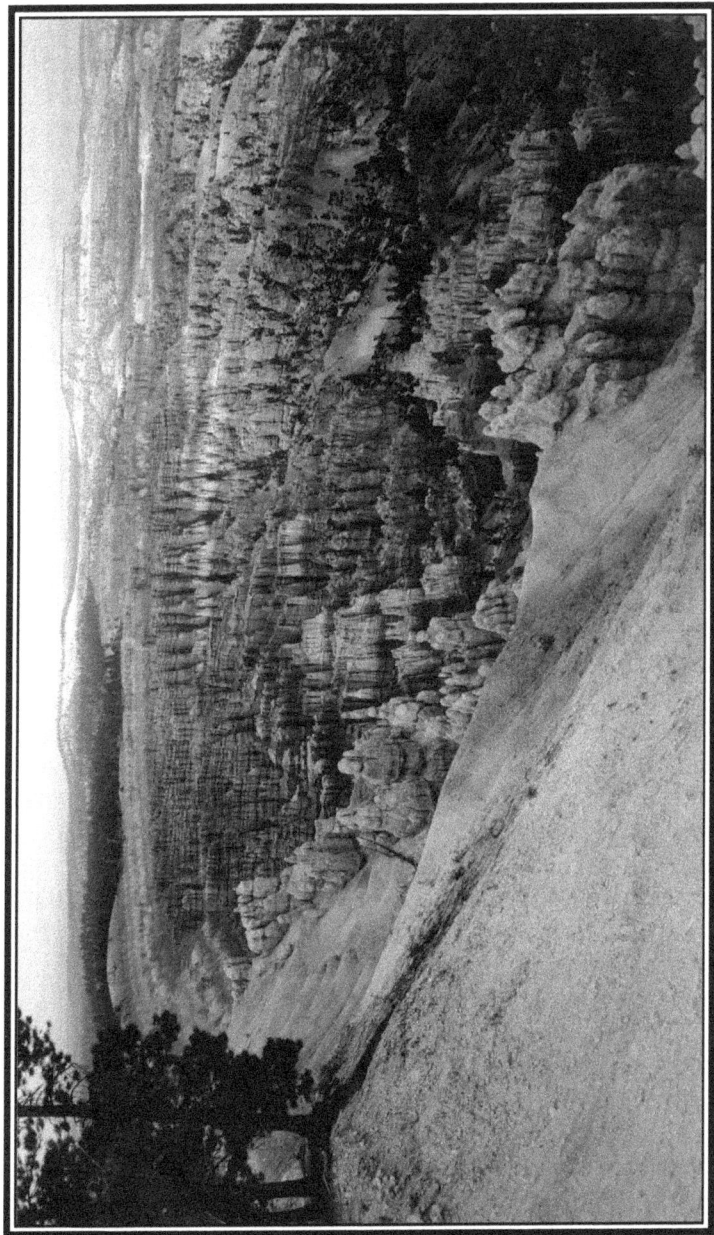

Bryce Canyon, Southern Utah. Photograph by Shipler.

It would take a long epoch for a thousand feet of lake sediment to accumulate. These sediments must have been accumulating untold ages before the coming of Bonneville. Their varying character suggests the deposits of alternating periods of moisture and drought. Evidently several fossil lakes, one above the other, were buried beneath these layers of salty sediments. Eight ruined, buried Troys are beneath the existing city of Troy. Evidently, many fossil lake beds, including Bonneville the first, are buried with their salt, soda, and borax deposits beneath Salt Lake. But these vast salty deposits of ages—of many fossil lakes—are overlaid with cemented, watertight sediments of rock and shale.

The Great Basin was neither round nor smooth bottomed. Its rim was broken and its bottom roughened with mountain peaks and canyons.

The lowlands of the Basin are from 4,000 to 5,000 feet above sea level, and the mountains rise 3,000 to 7,000 feet higher. The mountain summits have more precipitation than the lowlands, perhaps more than three times, and snow, during the winter, lies sometimes deeply.

A mountain revolution which began, perhaps, three million years ago made vast changes in the surfaces of the West. The Sierras were uplifted, numerous large areas sank, and the Wasatch Mountains were upraised. This revolution produced the Great Basin, an area between the Wasatch and the Sierra about the size of France. Rivers poured into this basin, but for ages it did not send a single drop of water to the sea.

Embraced within the Great Basin are parts of Utah, Nevada, California, Wyoming, Idaho, and Oregon. It is the result of a mountain revolution which extended over a long age of time during which blocks of the basin were uplifted and adjacent ones subsided. The Sierras and the Wasatch were the great boundary uplifts. This shut the drainage off from the sea and nearly cut off incoming moisture-laden clouds.

After two explorations of the region, I went away planning to return and follow the Provo shoreline from Ogden to the old outlet of the lake. This would take me

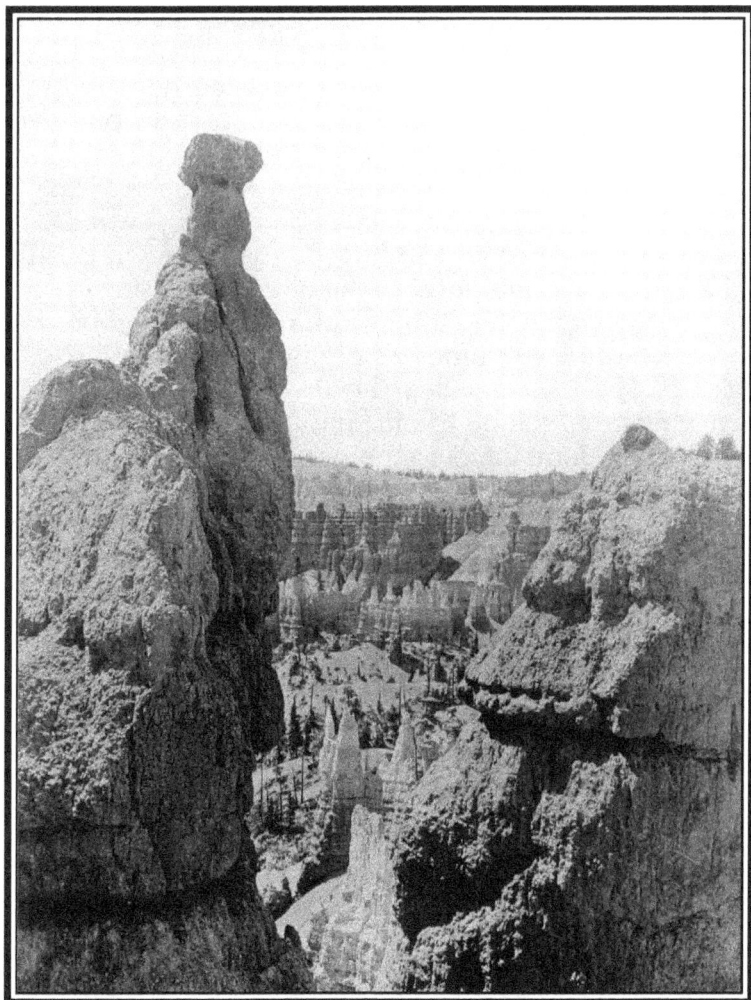

Temples of the Gods, Sevier National Forest
(now Dixie National Forest), Utah.

through a long-settled and populated section of the old Basin.

Meantime, I wrote for information concerning it. A governor took pains to explain to me that Bonneville could not have emptied into Snake River, as this would require it to run uphill. "I financed an irrigation ditch," he wrote, "which the engineering built for the water to run up grade, and it would not work." A few days later, he wired me, "Bonneville did it once and has been a dead one ever since."

At last I went to investigate for myself. Before leaving Ogden, I spent a day on Observation Peak, a few miles to the southwest. I could see into three or four states, and what was better, into most of the old lake's basin.

I could outline the shorelines and the islands of present Salt Lake and trace hundreds of miles of the fossil shorelines of Lake Bonneville. The Provo shoreline showed distinctly on Antelope Island, and it also came out on the Promontory Mountains to the northwest. Below and just before me were two vast deltas against the bold Provo shoreline at Ogden. In places, these enormous overlapping deltas, roughly level on top, outreached six to eight miles into the old lake. One section, deeply channeled by a river since the lowering of Bonneville below Provo level, was occupied by a railroad. A part of Ogden sat serene on one shoreline terrace and acres of cherry orchards were rooted in the rich delta sediments. Many miles of stretches of these old shorelines now are automobile roads.

The entire ninety-mile walk along the Provo shoreline between Ogden and the old outlet in Idaho was over the old lake bed and in places four hundred feet below the Bonneville level. The Oregon Short Line is in this lake bottom. Several times I wished there had been children with me to enjoy the romance of it, for none of the older people along the way caught the story.

I passed a plant that was making brick out of clay from the old lake bed; saw hundreds of cattle grazing the grass-grown sediments and hundreds of acres of peach trees upon old deltas. Thousands of Utah's farms are in the old lake bed. In places I saw three distinct shorelines on the mountains above us, and in Boxelder Canyon the lines were on both sides of the canyon. At several places, the shorelines folded back into a canyon, and across on the mountain wall of a narrow valley I saw the Provo, which I was following, the Bonneville nearly four hundred feet above me, and four or five less obvious lines between.

The region had been long populated and prosperous. Local people were ever pleasant, and a miner reminded me that a number of rich Utah mines were below the level of the old lake.

Everywhere, during these lake explorations, I had interesting glimpses into human nature—and our school system.

Near the Idaho line, I came to a country school that stood in an extensive plain, evidently the level sediments of the old lake bed. It was noon, and the teacher, a college man of forty years, and the collected pupils revealed no interest in my fossil geology, and when I pointed out before them a butte, broadly, boldly belted with an old shoreline, that had been an island in Lake Bonneville, I found that their interest, like in many sections of the Provo line, was absent.

Around Red Rock Pass, by the old outlet of Bonneville, I found the people most responsive, but completely oblivious of the stirring geological scenes that have been shown in their mountain valley.

Across the Pass, a railroad runs for a few miles through the old outlet channel of the Bonneville River. Immediately on arriving in the Pass, I climbed from the Provo up to the Bonneville at the summit where the water had first flowed over. From this place and level, the outrushing Bonneville River had speedily eroded down to solid limestone 375 feet below. After tracing the Bonneville line to its farthest point, I descended to the Provo and asked questions of the railroad agent. He had been stationed there a number of years and had not heard of it, and felt that if there ever had been such an outlet that it must have been elsewhere. Other local people along the miles of outlet visited had not heard of it, and only one had heard of Lake Bonneville. A few even remarked that there might recently have been an outlet of the state asylum. Still, this lack of interest in local geography is common, as I found when looking over the glaciology of the Park section of Manhattan Island and asking local people concerning the morainal story of Long Island.

The Red Rock Pass Region is rich in geological stories. Through this pass, perhaps for a few thousand years, poured the only running water that has come from the Great Basin, and this basin may have been in existence more than a million years.

This pass, when the eager waters of Bonneville hurried to escape across it, was deeply covered with loose, gravelly sediment washed down from the heights. This loose material was quickly washed away, and the outgoing Bonneville River speedily cut down three hundred and seventy-five feet. Then, dense, durable limestone was struck. This held, at this level the lake cut the deep, broad, and well-marked Provo shoreline, inside and against which incoming streams accumulated deltas.

The lowering of this pass removed its steep approaches and changed it into a mountain valley. The down-cutting of the river shifted the summit of the pass to Swan Lake, seven miles to the south of Red Rock. The broad, deep channel or valley worn by Bonneville River now seems like an abandoned channel, so small is the stream—Marsh Creek—which uses it.

In places it is dammed, or nearly so, and for miles along its moderate slope there are ponds and marshes. This stream, on reaching the nearly level grade through the pass, drops much of the carried sediment. In time, a delta-like dam is formed. This temporarily blocks the flow. As a result, at times Marsh Creek flows down into Bear Creek, and into old Bonneville basin; at other times, down its regular channel to the north to Snake River.

The wide old channel of the Bonneville River is one hundred feet below the level of a lava plateau. Perhaps twenty miles below the summit, the Portneuf River breaks into this counter-sunk valley and flows down this east side of it. In the west side of the valley is Marsh Creek. This is separated from Portneuf by a narrow tongue of lava.

Bryce Canyon, Southern Utah. Photograph by Shipler.

After several smiles of paralleling, these streams unite and empty, where Bonneville briefly poured, into the Snake River near Pocatello, Idaho.

Looking Glass Rock, Southwest of La Sal Mountains, Utah. Photo by Whitman Cross.

Rivers as Scene Shifters

An old Indian legend has it that all the river canyons of the world were made by lightning. This was a striking enough explanation, but the stirring facts of geology show that even the most radical changes in the face of the earth were not brought about by lightning-like changes.

On one of my later visits to Niagra, I had the good fortune to climb into a cab with the celebrated geologist, Professor Ralph S. Tarr. The cab driver was an expert in commenting faker-fashion upon the phenomena of the region, but had not been brought up properly on facts. When two or three miles down the river, we dismissed the driver and strolled back. It was a walk full of discussions of the Ice Age, the history of the Great Lakes, the origin of the Falls, and the adventurous biography of the Niagara River.

Once upon a time, three of the lakes, Superior, Michigan, and Huron, were drained by the Ottawa River, which took a short cut from Huron to the St. Lawrence. The drainage of the lakes appears to have once been to the Atlantic through the Mohawk and the Hudson; and at an earlier time through the Wabash by Fort Wayne into the Mississippi. The Niagara River is new, a recent outlet of the Great Lakes, and over most of its length has cut down only a few feet into the surface of the tableland. And the river could say with Mark Twin: "Boys, the Falls are a success."

Erosion is continually shaping new landscapes. It has made hundreds of canyons and thousands of vast valleys the earth over. There are no eternal hills or mountains. Mountains and plains, hills and canyons, deposits of iron, coal, gold, granite, soil, and sand are alike cut to pieces and moved away. Erosive forces break up the earth's surface at one place and carry it to another.

Chiefly under the influence of water, valleys grow wider, deeper, and longer. At the upper end, back-cutting slowly lengthens them. A valley often taps and absorbs another valley. Sometimes two valleys with a watershed or hill between erode and remove the barrier and become

"Speedy streams help dig with tools of sand and stones
the countless channels that we know."

one valley.

Two pirate rivers in the Andes displaced the Continental Divide and came near to starting an international war. The boundary line between Chile and the Argentine Republic originally followed the Continental Divide. This was assumed to be permanent. But it was not. The steep Chilean rivers, ignoring the old topography, committed swift and picturesque piracy. Eroding rapidly headward with flank advance, they beheaded Argentine rivers and compelled the waters of these to flow to the Pacific. They broke and displaced natural and national boundaries.

Streams struggled for existence and sometimes break through an established stream boundary and invade headwaters of an adjacent stream. The invader geologically is called a "pirate," and the stream seized is said to be "beheaded." The international commission selected to settle the disputed boundary line found that no map had ever been made of it. The survey then made revealed an interesting river habit—piracy; which included invasion, seizure, river adventure, and a revised geography lesson.

The shifting channel of the lower Rio Grande is about as permanent as a mirage, and the bank and island residents of this stream may wake up any morning and have topographic allegiance to the nation on the other side of the river. It is difficult to say whether this kind of stream is a born adventurer or is endeavoring to make a permanent peace commission necessary.

Stupendous are the changes wrought by running water, as the Grand Canyon shows. Every stream is a landscape artist, a continental sculptor, an erosive power of first magnitude. The earth's surface is being steadily lowered by the aged and endless activity of rivers. Old surfaces and scenery are melting into the new before the ever-changing rivers and the changeless sun. The artistic relief work of streams—hills and dales, countless cliffs and canyons, valleys and plains—is all about us.

The struggles of streams in hills and mountains appear almost in the nature of feuds decades long. Territory and channels are taken and retaken. These feuds have oscillated watersheds on the map and resulted in that

Athabaska Falls Canyon.

determining topographic power known as geographic environment. The earth channels created by streams have directed his development even hundreds of miles from their original scenes and ages after they ceased to play at path making.

In the Adirondacks, there were contending Indian tribes who never quite settled a boundary line. And amid these poetic and geologically ancient mountains old river channels appear to be about nine deep. The Potomac River heads in behind a mountain range and owes its present prestige to past piracy. The Appalachian and Blue Ridge mountains are full of graphic records of river piracy and mountain stream feuds. These contending streams cut channels and gouged new gaps without the slightest regard to the old regime, and with no thought for innocent by-standers or new geography lessons. But their audacious and influential piracy contributed something to posterity. Cumberland Gap, that strategic thoroughfare of both peace and war, through which the westward course of empire took its way, is the channel of a beheaded stream, a base level but through a mountain barrier abandoned by the pirate.

All rivers lengthen by cutting headward. Stream piracy may result from advantages possessed by one stream over its competitor—steeper grades and softer rocks, which enable it to cut its way more rapidly, to lengthen its lines, to seize and intrench itself successfully beyond and behind the neighboring stream, which has to cut through resistant rock with slow grades. However, the pirate often has his day and is then outpirated. An obsolete stream on the other side of the top cuts through into a stratum that gives easy going; or warping of the surface gives it high speed and all the power for piracy.

The geological story of the Tennessee River is one of rare romance and piratical adventure, one of the best in river biography. This stream of mighty flow, and now deeply intrenched main line of communication, has had a turbulent and contentious past. Parts of its stirring records have been washed out; others, time has only camouflaged with forest, vine, and flower. The Tennessee River is made up of three ancient, once-honored rivers which formerly flowed in other channels, each over it individual course and toward a different point of the compass. It is a union of streams formerly independent, and of watersheds which once had different connections and allegiances.

Sometimes, after a stream has lost out and become old and decrepit, a sudden upheaval or a soft geological stratum occurs—and old landmarks are swept away, and the long-fixed watershed conditions are changed.

The middle courses of the river usually are safely intrenched for long periods, but if it reach the sea, the ever-lengthening lower course that the stream builds for itself is over a channel subject to breaks, bendings, and abandonment.

Streams meet with accidents. The sinking of the territory across which a stream flows causes the valley to be drowned. The region around Chesapeake Bay has sunk and submerged the lower ends of a number of streams. Sometimes, earth tilting gives a slow stream a steeper grade. This it accepts merrily and at once makes use of in erosion.

There is severe competition in the animal world. Species crowd species and endlessly contends for place and food; but at times there is a truce. Often they co-operate, frequently a species forms alliance with a foreign one and with this fights a common enemy. Much of their advance is due to cooperation and mutual aid.

But a stream's source seems never to be settled. The war with its competitors does not end. Its competitor suddenly speeds up, due to local uplift or easier excavating, and seizes the entire summit. After centuries of successes, when a stream source has almost secured all the coveted watershed and has dug itself in—is deeply entrenched—then the current changes and it loses. It may sink into insignificance or even be completely absorbed and eliminated from the map.

Rivers, like glaciers, make vast and picturesque excavations headward—"backward," as an old prospector put it. By this apparent working backward, by advancing in one direction while flowing in the opposite, they remove

"Some guides will work in the vast canyons. Royal Gorge from the summit, Colorado."

hills and sometimes dig a channel through a mountain barrier.

Excessive saturation, soft strata on steep slopes, causes earthworks to slip and to move as landslides. Landslides and snow slides carry the heights and slopes down into the lowlands. A million tons or more may slide at one time. Many lakes are formed by landslides damming a waterway. Or the debris forms dams, more or less permanent, in river channels, temporarily overwhelming the stream with sediment to transport. The Cascades in the Columbia, the Rapids in the Colorado, are in part wrecked peaks and cliffs from a former place in the sun. A landslide often blocks a channel and compels the river to seek or make a new course.

Yellowstone Lake for ages sent its waters to the Pacific. But something, not many centuries ago, appears to have blocked the old outlet. The water proceeded to add interesting revision to the geography of this wonderland; they abandoned an old and added a new section of Continental Divide and cut a channel that gave connections with the Atlantic. This added the falls and the Golden Canyon with sunset colors—a new and shining landscape.

Running water thus is the mighty artistic director of the earth. With its ever-active rivers, it is constantly making scenes and constantly changing or merging them into new ones. The topography of any section is geologically described as "being born, going through periods of youth, decline, and old age." It is changed. A river is the topographic scene shifter.

Rivers have their day. They have youth, growth, decline, and old age. A river has an adventurous life, so much so that geologists, who use mostly mathematical words, tell of a river meeting with "accident," being "beheaded," "drowned," and "struggling for existence." Many a river has a biography well set with excitement and dramatic scene.

Countless fine-toothed streams are constantly combing and raking the entire land surface of the globe. Flying squadrons of clouds fill the sky and drop untold tons of tiny water bombs upon the earth. Gravity directs the

Panoramic View of Colorado National Monument. Photograph by George L. Beam.

gathering and transportation of the wreckage. Frost and chemistry are also diggers—the forces used to break, loosen, and dissolve the surface. The fluency of water in connection with gravity, evaporation, and precipitation gives it alliances and perpetual motion for transporting the materials gathered from the surface of all continents. Speedy streams help dig with tools of sand and stones the countless channels that we know.

The sea, sending its messages by cloud and storm, lays tribute on all the land that lies above the surface of the waves. Each little drop of water conducts a little grain of sand back to the sea, and with the cooperation of chemistry, each drop also carries salt and lime and phosphate in solution for sea gardens, coral building, and the flavor of the sea.

Beaver ponds are constantly silting up with sea-bound sediment, hampering erosion and stream success. Millions of these have filled, reduced sediment delivery, and added empires of rich soil, changed boundaries of life zones, and invited meadows and forest to expand. But erosion never ceases.

Sediment and sand washed into and completely filled the old pond where, as a boy, I had watched beavers play. With difficulty, I traced the almost buried grass-grown dam. On top of the ruined house, several small pines were growing among gentians, orchids, and tiger lilies. A number of willow clumps, a small grove of aspens, and a few tiny spruce and fir trees were growing in the soil where the pond used to be. The site was covered with grass; the pond where the beavers had lived and played was filled in, overgrown and forgotten. The pond was on a stream that came down from Long's Peak along the top of a high moraine. In places, the pond had been four feet deep.

Each year, the rivers of the United States alone mine and move 513,000,000 tons of solid matter in sediment and 270,000,000 tons of land surface in solution back into the waters of the sea.

In boating down the Mississippi years ago I landed on an island where I met a man who explained to me the

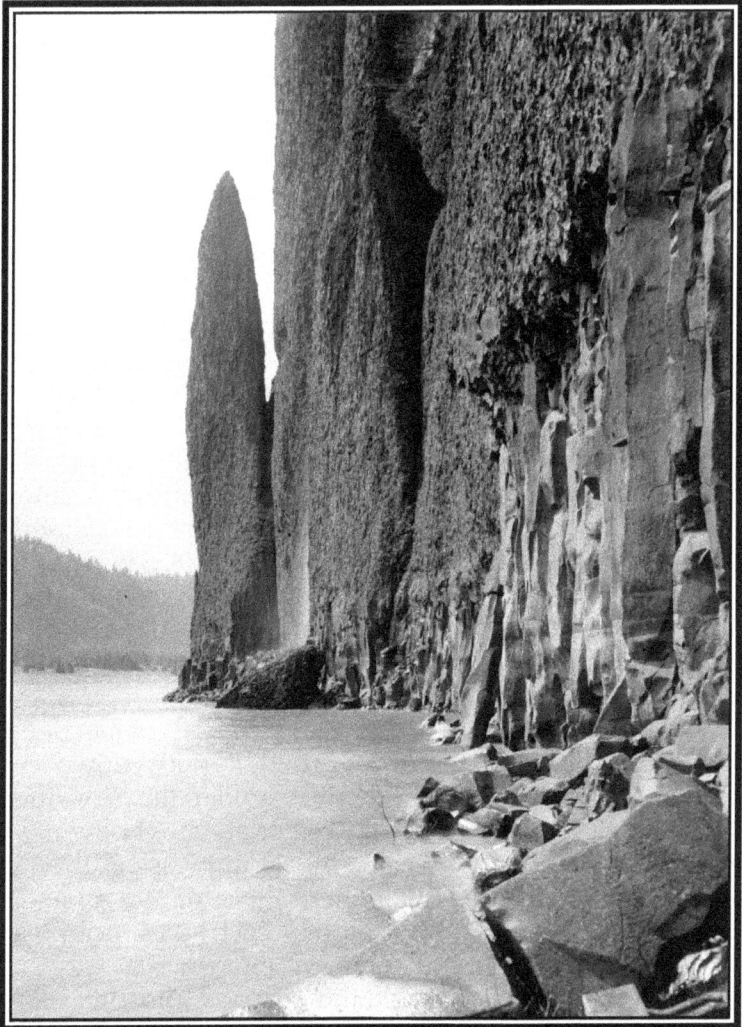

"Cigar Rock, Columbia River, 25 miles east of Portland, Oregon, showing more of Nature's handiwork in chiseling out this peculiar formation for our amusement and study. It towers many hundred feet above our heads, and serves as a buoy marking the deep channel in the river." Photograph by Weister Co., Portland, Oregon.

island's origin. When a young man, he was going down the river and his steamer struck a snag and sank. A sand bar quickly formed in the eddy. It is well known that many sand bars form around a mere snag in a river. Unable to rescue his sunken boat the man took up his residence on this sand bar which had grown into the island we were on.

"Have you ever thought," I asked him, "how much lime solution from Kentucky caves is annually carried into the sea?"

"Enough to give the present oyster population a three years' supply," was his instant reply.

The old ever-muddy Missouri River still flows on. It is heavily laden with liquid land taken from everywhere which will be piled somewhere. In this dull river the imagination sees fertile tree-fringed farms across which happy people move, and delta landscapes where the restless ocean washes mysterious and adventurous sands.

A delta forms wherever a stream deposits its sediment—in a lake, another stream, or the sea. A delta may be a tiny affair in a brook, a little fairy playground. Many deltas are of enormous size, compromising thousands of square miles. The delta of the Merced River in California is forty miles across. The surface area gives no idea of the bulk of the deltas; many of them are hundreds of feet deep.

Every stream is a scene shifter: it pulls down old landscapes, carries them far away under the sky, then readjusts and recasts this old material. A wholly new, substantial landscape is given its impressive though transient place in the white light of the sun. The Amazon, the greatest river on the globe, each day pours an inconceivable quantity of sediment into the sea. Day and night, ever flowing through the seasons and the years, many mighty rivers are pouring their loads and building enormous deltas out into the seven seas. Slowly the earth is becoming lower and wider.

The seas with surge, current, and breakers ever wage their endless roaring war on these encroachments. But on many sectors of the battle front of sea and land, the land has made the effective drive. That shifting ally of land and water, the wind, generally aids the sea. The wind tears the earth, erodes the surface, and blows dust and sand afar. The amount which it annually moves equals millions of tons. Along the sea and lake shores, by thousands of streams, in deserts and plains, the wind is almost continually scattering, moving, and piling sand. Sometimes it is heaped into dunes. In the West, dunes occupy thousands of square miles. They are found in Nebraska, Kansas, Nevada, and Arizona, and along the Columbia River.

A continuous stream flow pours forward a never-ending supply of sediment reserves which gravity eagerly, steadily conscripts from thousands of square miles of surface; and these reserves of little grains of sand are so strategically, steadily advanced that the little drops of water bend back. On every shore, the deltas are pushing forward the waters of the sea.

Rivers on every coast continue to build beyond the shoreline. Sometime there will be soil and sediment far out in the ocean—leagues of deltas throbbing with many kinds of life where now sharks swim amid the surges of the sea. Thus deltas rise as new landscapes on the shore of the restless sea—landscapes that formerly were part of the hills and mountains on a far-off skyline.

Wind and water, birds and animals, carry seeds of grass, trees, and flowers to the fresh soil beds—the new landscapes. Numerous forests and farms, magnificent and home-filled distances are scattered along the delta shores where formerly waves of the seas and waters of lakes rolled or rested.

Placer mining is mostly carried on in deposits of eroded material, deposits formed by wind, glaciers, or water. The prospector with his gold pan has gone around the world panning the deposits along stream and shore. Occasionally, he has found the long-sought gold in the midst of a delta.

Numerous cities of prominence stand upon deltas that streams built during the ages of the past. Interlaken, Switzerland, is upon a lake delta. Hwang-ho delta has built up three hundred and fifty miles. In 1851 the Hwang-ho River, instead of emptying into the Yellow Sea, shifted to the Gulf of Pechili, three hundred miles to the north. As this was between dikes with a surface level fifteen to thirty feet above most street and farm levels, its change was disturbing.

New Orleans, now a considerable journey from the mouth of the Mississippi, is built wholly upon a delta, and so low-lying is it and so built up is the river bed, that the traveler climbs flights of stairs to get aboard the steamer.

Each day the mighty Mississippi pours more than a million tons of earth sediment and solution into the Gulf. The delta of this river is each day advancing its ragged, sectored shoreline nearly one foot into the Gulf, advancing about one mile every sixteen years. In the course of a generation or two, this ever-growing land area will compare favorably in size with some of the European nations that are struggling for a place in the sun.

The Mississippi River delta contains age-old wreckage; it is a continental contribution built by the Father of Waters. It is a mingling of mountain fragments and broken farms, the blended ruin and richness of ten thousand plains and peaks. In it, side by side, lie remnants of Pikes Peak, the heart of old Kentucky, a part of the Mammoth Cave, lava from old Yellowstone fires, glacial silt from Canadian mountains, dust from the Great Plains, sediments from rocks that were formed in ancient seas, and even the black meteoric dust of burnt-out worlds and stars. A delta may be a combination of all geological rock strata and of all life that has lived its little day and returned to dust, and may carry even the wreckage of other worlds than ours.

A polished piece of granite in this delta may be as old, almost, as the earth. Erosion on Canadian mountains unearthed it. The southward sweep of the Ice King seized it, carried it a thousand miles southward, grinding and reducing it, then depositing it in Ohio. Here a flood seized it, rushed it to a sand bar in the Mississippi River, and it lingered. By slow stages it rolled its way down the Mississippi channel and at last came to rest within sound of the sea.

Sections of this cosmopolitan delta may be submerged in Nature's melting pot, become a stratum, lie buried and forgotten for ages beneath the sea, then reappear in fiery volcano or uplifted stratum, in new landscape slowly

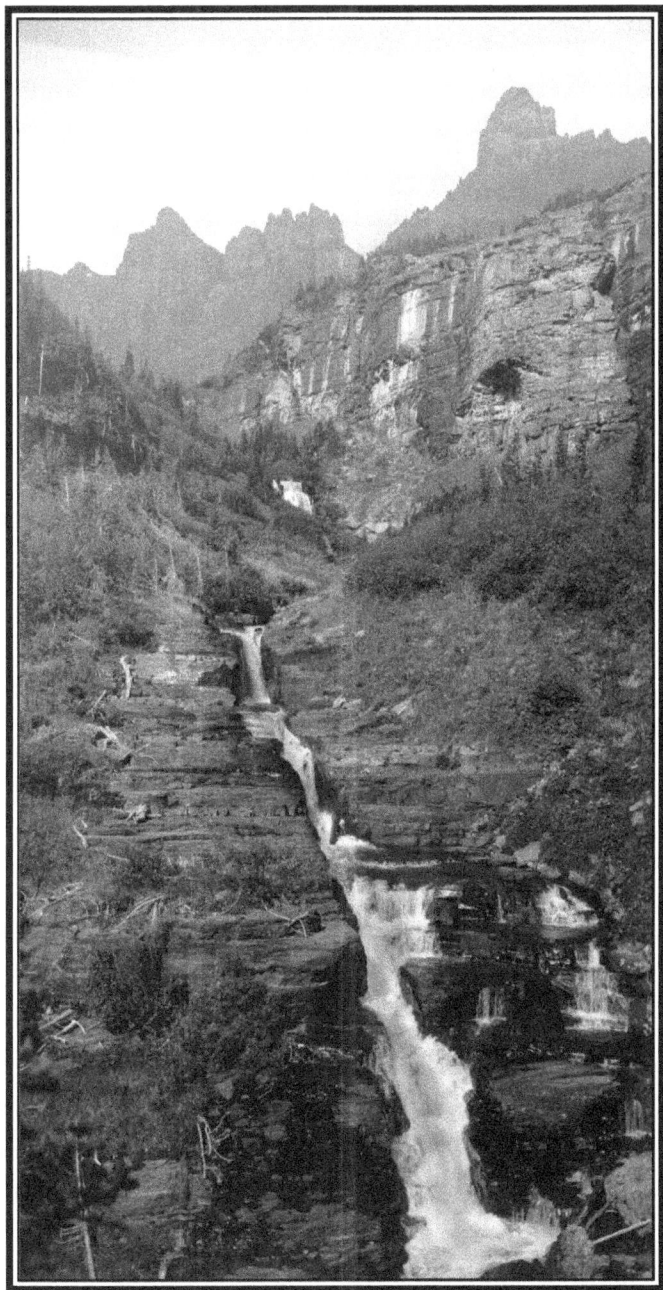

Golden Stair Falls, just below Iceberg Lake,
Glacier National Park. Photograph by Frank
Kiser, for Great Northern Railway, 1910.

wasting under the erosive forces of unmeasured time.

Scientists tell us that this continent is being lowered at the rate of one foot every nine thousand years. Continents have been uplifted and worn down again and again. Geologists say that the sedimentary rock strata of the earth have a total thickness of forty miles. But we are in these scenes such a little while—make a mere transient visit—that we cannot see, except with revealing mathematics and poetry, that the most substantial landscapes and age-wrought scenes are moving pictures—moving through a vast orbit of endless change.

Erosion showed monumental results during the Mesozoic era. The earth entered this era with mountains towering on all the continents in the Northern Hemisphere and the rivers changing their landscapes. The rivers, though acting leisurely, were ever eroding and made changes vast. Mountain horizons were peopled with ponderous and ancient life; the earth circled away in alternating day and night. During its countless and advancing changes in earth history, the rivers ran on. Before the day was done—before the Mesozoic era had given way to its successor—the mountains which, over the Northern Hemisphere, had greeted its morning had been worn down by running water; but the rivers now were slow and sluggish and the landscapes in which they had a place in America, Europe, and Asia were but vast low plains.

Landscapes are being absorbed. Every scene-shifting stream has salt for the sea, lime for shells that will strew beaches yet unborn and for coral islands on whose white and palm-plumed shores deep blue oceans yet shall roll; sediment and chemicals with which to make and color the strata that are yet to rise from ocean's depth and have a place—be a crumbling and transient landscape in the sun.

Launching Icebergs

One May, more than a quarter of a century ago, a whaling vessel lowered a boat, two Indians, and myself on the Alaskan coast supposedly by the entrance of the Muir Inlet. Rowing inland, we broke abruptly through the fog screen into the midst of a fleet of icebergs. Many were of stupendous size and several were of striking ice architecture. One pinnacled berg appeared like an enormous five-master. A majority of this strange fleet shone dazzlingly white in the morning sun, with blue-black shadows. There were stragglers, gray-black like colliers, and a few scattered ones of marvelous blue.

We pushed up the bay and presently were pulling to right and left among the icebergs putting out to sea, watching on our left the broken, bristling ice cliffs—the fronts of glaciers—against which the waves were washing. Occasionally, a heavy, towering mass of ice collapsed, creating terrific explosions in the water and sending rings of violent waves rushing toward every part of the bay. There was an almost continuous roar and splash of these heavy waves as they dashed upon the countless bergs scattered through the bay, causing them to rise and roll long after the wave had collapsed high up on miles of distant broken shore.

The Indians, munching fish eggs, watched the strange moving exhibit with interest, but fortunately with less enthusiasm than myself. Two heavy swells from launched icebergs rushed our boat and nearly spilled us as we swished over the top. The Indians insisted on our keeping about a quarter of a mile distant from cliff fronts, where bergs were launched and storm waves started.

However, we were caught by a danger unsuspected by the Indians and to me unheard-of. We were headed for a distant inland channel, and several times dashed between close-drifting bergs that threatened to crush us. We watched that these did not bow a shattered pinnacle upon us or that their falling ice chunks and boulders did not explode and deluge us with small fountains.

At last, we came into a stretch of open water. Not a wave was in sight, and a solitary big berg near us appeared asleep. Suddenly, we were lifted into the air upon upraised water, and for a moment looked down upon the top of this big berg. An enormous blue ice mass had broken loose from the depths and risen under our boat. Then we were swished shoreward on a wild, high wave, which flung us out of the bay.

We dragged our drenched selves from an alder thicket sixty feet above the shoreline. One of the Indians was still munching dried fish eggs. The alder clumps had been our shock absorber, but the boat had broken its head against one boulder and its back across another. Dripping, we three stood for a moment watching all our food and bedding floating off with the flotsam and jetsam of the bay.

The boat was smashed, the outfits a total loss; but flopping among the willows and alders were hundreds of fish, which were flung ashore by the wave that changed us to castaways. We built a driftwood fire among the alders and boulders, and as we steamed, we looked in and round the bay upon one of the grandest glacial exhibits in all the world. We had missed Muir Inlet, but had landed in the unrivaled Yakutat Bay.

The detached iceberg that wrecked us had risen from the bottom of the bay a thousand feet in advance of the visible front of the glacier. This submarine berg was a deep blue, but changed rapidly to white.

A number of the many glaciers that terminated in the bay were sliding in canyon channels which bottom a few hundred feet below water level, while the tops of their ice fronts stood two hundred feet above the water. That part above water level was cut off by wave action and detached as icebergs more rapidly than the submerged invisible part. Apparently, all blue bergs rose from the depths, and these changed rapidly to white. The gray-black bergs were masses of glacial debris—gravel and boulders.

This mountain-locked harbor appeared to contain all the glaciers and icebergs of creation. The mountain walls were so thickly, heavily laden with ice and snow that the

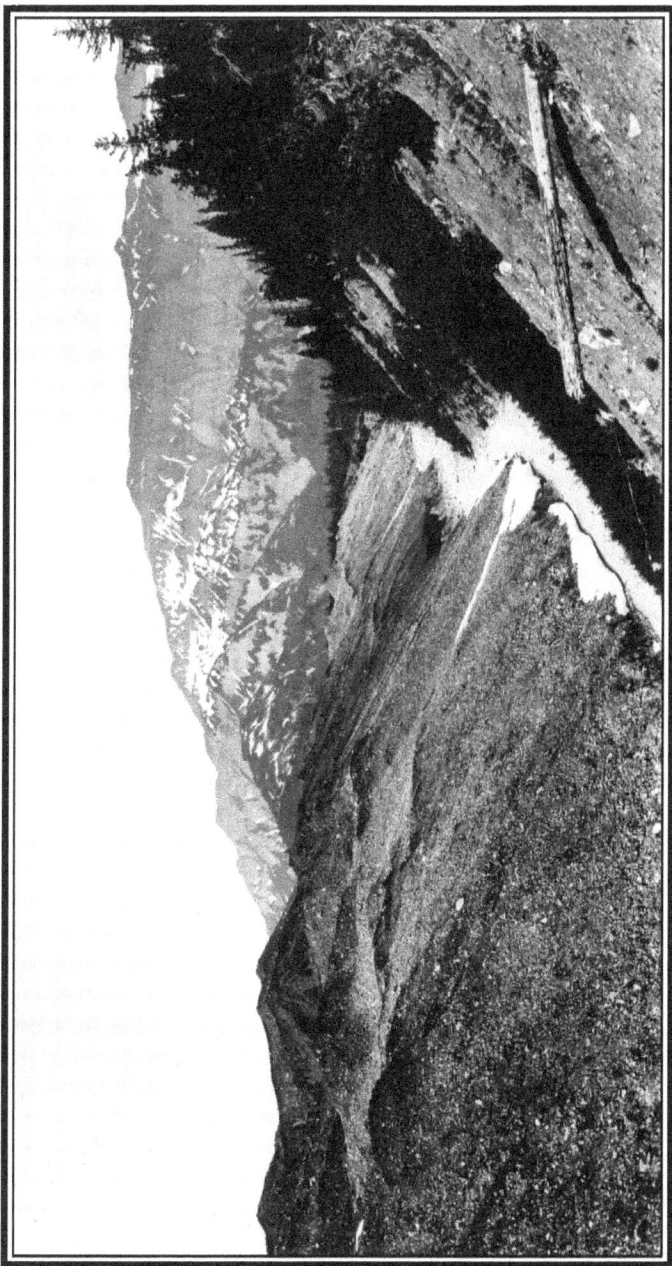

Marginal drainage at the southern base of Chaix Hills, Alaska, looking west. Photograph by Russell.

rocks were only here and there visible. The adjacent white mountains send down mile-wide glaciers which terminate in this bay; launch ships of white—icebergs—which later go down to the sea.

I.C. Russel, the celebrated geologist and glaciologist, had explored this scene a year earlier, and Frederick Funston had landed somewhere in the region only a few days before me.

I was bound for the interior of Alaska, but thought to visit the Muir Glacier, in which Muir had interested me, while waiting for the excess of snow to clear from the Chilcoot Pass trail. My plan was to repair the broken boat and with this go for another and supplies. These could perhaps be obtained at the nearest Indian encampment. The two Indians said that with repair materials they could put the humpty-dumpty boat together again. All the remainder of the day, we three searched miles of shoreline among the boulders and alders, and that evening had a pile of fragments—broken boxes and their precious nails, rope, a few tin cans, and the green and invaluable skin of a wolf that had evidently been killed by a wave rush which crushed him against the boulders.

We broiled fish for supper and lay down without bedding between driftwood fires. The night was still except for the falling ice cliffs and the wash from their waves. The stars were near, and the snowy mountains made splendid marble architecture in the night.

Leaving the Indians struggling with the broken fragments of the boat, I next morning climbed a high, commanding point above the bay. Snowy mountains, glaciers, and icy peninsulas edged the water. Everything was on a stupendous scale. A wide canyon below me carried a glacier that extended miles and leagues back into the high white mountains. A snow slide gave an excellent exhibition by plunging down upon the glacier. The slide was so far away that I heard not a sound, but so large was it that its lurches, leaps, and curvings were easily seen. A thousand-foot column of agitated snow dust rolled up and stood briefly over its roughened mass, where this stopped half a mile out on the glacier.

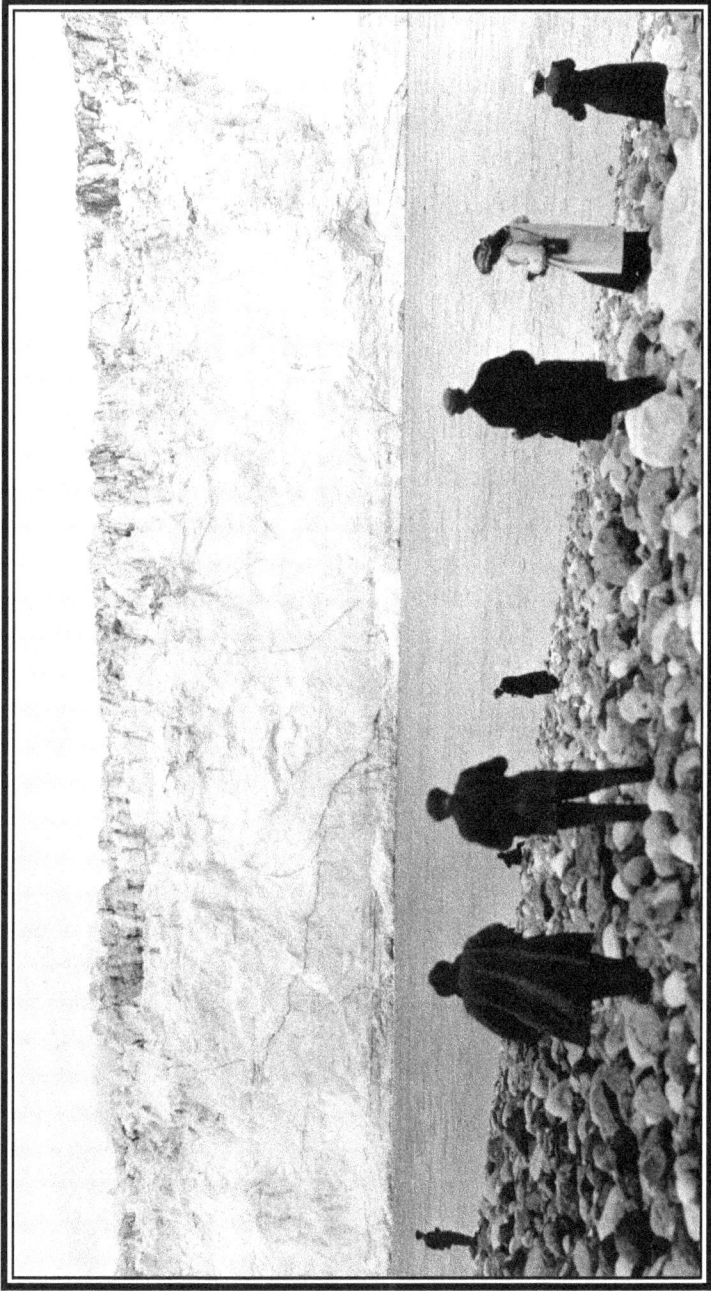

People at Child's Glacier, Alaska. Photographer unknown.

Child's Glacier, near Cordova, Alaska.
Photographer unknown.

One avalanche, a mixture of rocks, ice, and snow, started near me and crashed down upon the glacier. For longer than a minute, its echoes and reechoes rioted so vigorously among the snowy cliffs and icy canyons that I looked, expecting to see something in action. When the avalanche came to a stop out on the ice, the mass appeared as large as several *Mauretanias*. So extensive was the scene that, when I lowered my field glass, I had difficulty in finding it with my good eyes.

Northward, as far as the eye could reach, was a vast desert of snow. Many mountains appeared to be made of it; others were deeply buried beneath it; and here and there the tip of a peak barely pierced its heavy stratum.

What an array of water in cold storage! A snow desert as large as two or three New England states, together with hundreds of square miles of ice. In due time, all this crystal cloud material would be shaped into finished products—icebergs. These would be launched by the glaciers, exhibited in the bay in front of the steep white mountains, then sent forth on a strange sea voyage to melt and mingle again with the waves and clouds.

Off in the distant west lay what I took to be the Malispina Glacier. It occupied an empire of surface and was so nearly stagnant that groves were growing in its debris-covered back.

The 2,000-mile stretch of Pacific Coast between the mouth of the Columbia River and Cooks Inlet, Alaska, has an extremely heavy snowfall: sixty feet a year and up-ward—mostly upward. The Yakutat Bay-Mount St. Elias region is laden beneath its full-heaped share. More than snow falls each year than melts. The accumulating snow quickly changes to ice through compression and partial melting. As this ice mass becomes sufficiently weighty, it begins to crawl down slopes. It becomes a flowing ice river—a glacier.

Glaciers, like water rivers, move forward along the line of least resistance. The rate of movement depends on the weight of the mass, the degree of steepness and roughness of the slope down which it moves. Small and nearly stagnant glaciers advance from one to twelve inches a day, but the majority of glaciers go forward from one to ten feet a day. On rare occasions, a combination of favorable conditions may cause any glacier to lurch briefly and slide forward at greatly increased speed.

A few years ago, an earthquake in Alaska temporarily put new life into numerous glaciers. They were shaken out of slow-going ways. The Muir Glacier was shattered and changed by the quake. Its lower reaches slid forward and so jammed its terminal bay with icebergs that steamers were unable to enter the bay for two years. By the time the bergs had cleared, the end of the glacier had retreated and no long reached tidewater.

For a few days following the earthquake, a number of glaciers rushed ice deliveries—launched numbers of icebergs. This was followed by normal flow for some months; then an intensified, prolonged flow occurred, evidently due to the flood of glacial material—rocks, ice, and snow—which the earthquake had shaken down upon the source months before. These glaciers, one, two, and even three years after the quake, advanced, pushed their noses forward from a few feet to a quarter of a mile. This quake occurred a few years later than my visit.

I saw only one glacier that was advancing beyond its former terminus. It was one that melted away without reaching tidewater. Its thousand-foot front was ploughing through morainal deposits made years before. In places, this debris was nearly one hundred feet deep. Part of the moraine was covered with a spruce forest that was more than a century old. The crushed, cracking trees filled the air with the odor of balsam and pitch as the ponderous, irresistible mass pushed invisibly forward. In front of the ice mass, trees were leaning forward at every angle; numbers were uprooted, while others were down and the ice front sliding upon them. This forest now being flayed and crushed alive had grown in glacier-made soil—soil crushed and ground from rocks and distributed in other days by a glacier.

On the way back to camp, I walked two miles over the rough surface of the glacier on which I had seen the avalanche descend. One section evidently was above a

Southern face of Mt. St. Elias, Alaska. Photograph by Russell.

rough, steeply inclined place in the bottom of the channel—a place that would create wild rapids in a river. The slow-advancing ice opened into crevasses as it passed over this place. An enormous pile of rock debris that was emptying into these crevasses had slid down upon the glacier more than a mile upstream. The time required to advance this far had probably been about two years.

A goodly quantity of this rock-slide debris had already dropped into the yawning crevasses. While I stood near, several large rock fragments from the pile tumbled in, and on the caving edge, small stuff was almost constantly sliding or tumbling in. Down in the glacier, these rocks would be pressed powerfully together. Numbers probably would drop to the bottom, where the glacier, with a few hundred tons' pressure, would ride and slide upon them, crushing and grinding them against the bottom and each other as the ponderous glacier moved ever forward.

A glacier is a sculptor of the rock ball called the earth, and it carves the surface into canyons and plateaus, making scenery and soil. At the source of a glacier, as well as at crevasses, ice, snow, sand, gravel, and slide rock accumulate and mingle in the upper end of the channel, and this confused mass of cutting tools tears and polishes the sides and bottom of the canyon channel as the mass slides forward. Not only is the channel widened, deepened, and straightened, but the tools themselves are mostly worn to dust by the time the terminus, or end of the glacier, is reached.

The last Ice Age made vast changes in the topography of the Northern Hemisphere. It ground up and moved mountains, changed river channels, made thousands of lake basins and fiords and covered thousands of square miles with productive soil. Glaciers—compressed snow flowers—carve grand scenery and soil. Much of the soil in the temperate zone is largely made up of rock flour of glacial manufacture. The surface of several Mississippi Valley states is deeply overlaid with glacier grindings, and most forests in the Rockies and in the Cascades and the Sierras are standing in glacial soil.

Returning to camp after a long day among glacial

wonders, I found that serene Indians had made a start in assembling the fragments of our shattered boat. The repairs would require a few days longer to finish. As my assistance was declined, I took a hunk of broiled fish and set off for a two-day trip, hoping to reach the source of one of the glaciers. Among the willows by the lower end of the glacier near the bay, I found numbers of flocks of ptarmigan. A mile or so up the glacier on the south wall I saw a number of bighorn sheep.

This glacier was more than three miles wide and probably a thousand or more feet thick, and filled the bottom of a canyon from wall to wall. The snowy, icy walls rose, perhaps, two thousand feet higher. On top of the glacier, I walked eastward up this wild, white, wide avenue. The surface of the glacier, which appeared generally level, was mostly snow-covered. Most of the time, I was within a few hundred feet of the south wall, but kept this distance to be safe from falling rocks and down-rushing slides.

Another danger was from the snow-covered crevasses. Numbers of the wider and longer crevasses were either open or were separated by high, sharp ice ridges, which advertised the hidden dangers. In places, there would be a single narrow crevasse; in other places, half a hundred openings in close succession. Whenever there was any doubt, I explored with a long staff; but much of the time I was able to keep on the solid, snowless ice of wind-swept ridges, where there was no danger.

Mid-forenoon, a bear, evidently a grizzly, crossed the glacier from the south. I was in a hollow between snow-drifts and a crevasse, and he did not see me. When about a quarter of a mile out on the ice, he heard a snow slide behind and turned to watch it. This slide was closely followed by a rock slide, which went down with a thunderous roaring and crashing. The grizzly watched it, rising on hind feet. As soon as the straggling tail-end fragments ceased coming down, he went to the rock wreckage and climbed over it. Here and there, he stopped to eat something, probably roots. Leaving the wreckage, he followed his tracks back to the spot where he had stopped, turned for another look, then shuffled across to the north side, where he disappeared among the rocks.

Often I turned aside to examine the enormous piles of avalanche rocks that lay upon the glacier, and I came upon one that was thirty-two steps long. It was embedded slightly in the ice, but rose at least thirty feet above ice level. This enormous rock was floating down on the ice stream as readily as a chip floats on water. Of course, its progress was slow. It evidently had been carried about one mile.

On top of this wide glacial highway, I walked inland over hundreds of piles of debris, some almost pure snow, others mostly rocks and earth. The spring thaw evidently was the time of snow and rock slides, as the thaw was releasing the rocks wedged loose during the winter and loosening the big, steep-placed snowdrifts. As I could see miles ahead, with no end of the glacier in sight after six hours' walking, I turned aside to explore the source of a small tributary glacier or ice river.

Glaciers begin abruptly, like a river which starts in full volume from voluminous springs. This small glacier filled a tributary canyon about a mile long, which ended abruptly against a 1,000-foot wall. Down this wall and from slopes to right and left came snow slides and rock slides. A score or more of these had piled their contributions in one mass of fierce confusion a little below the uppermost end of this glacier. Rocks, ice, snow—in a pile four hundred or more feet high—were settling into place, and in a short time would be blended and a part of the slow-moving ice river.

Glaciers, like rivers, cut headward with surprising rapidly. The high, precipitous will in front of the head of this glacier evidently was due to the headward undermining and backcutting of the glacier. The crack, or *bergschrund*, which commonly is open between the upper end of the glacier and the snowfield or rock wall, allows air— and with it changing temperature—to reach beneath the upper end of the ice. This air and changing temperature means freezing and thawing, rapid rock disintegration and separation. Often, the upper end of the ice freezes fast to loosened blocks of rocks. These are then slowly dragged out. Long's Peak, Colorado, has been half carried away by the headward cutting of a glacier. This attacked its east wall from the abutting end of a glacier-filled canyon at an altitude of about twelve thousand feet, twenty-five hundred feet below the summit. In the Big Horn Mountains, Wyoming, are remnants of former peaks, the remainder having been carried slowly away by back-cutting glaciers. Canyons now are where peaks formerly stood.

Leaving this glacier-forming place, I started on the return journey, hoping to reach the coast before night. During the afternoon, I went across the glacier to examine a peninsula-like ridge of ice that thrust in a quarter of a mile from the north wall, and with a surface a few hundred feet higher than the general level of the surface of the glacier.

Evidently, there was an inthrusting rock ridge in the bottom of the canyon, and over this rock ridge, or peninsula, the glacier river flowed; for glaciers, like water under pressure, will flow up a grade, or uphill. The glacier was simply flowing up and over this inthrusting obstruction in its channel.

Sunset hour, with its long, ragged lights and shadows, was on the glacier when I left this deeply crevassed, icy peninsula and started on. It would require two hours to reach the coast, and as this could not be made before dark, I began to watch for a place to camp, as it would be perilous to travel among the glaciers in the dark.

Up on the north wall, several hundred feet above the glacier, was a grove of Sitka spruces. A part of this grove had been recently cut away by a snow slide. The trees thus wrecked lay before me in confusion on the ice. Many of the trees were smashed to cordwood, numbers were buried end-on several feet in the ice. On a bed of boughs, between two roaring fires, I had a fairly comfortable, primitive night.

The following day I spent among the glacier end in the edge of the bay, with its fleet of bergs. The bay is the launching harbor of many glaciers. One of these glaciers, then unnamed, thrust out into the bay an ice front that was at least four miles wide and with ice cliffs more than two

Crevasses on Arapaho Glacier, Rocky Mountain
National Park, September 1, 1911.

Photographs by Judge Junius Henderson.

hundred feet high. Two other glaciers were more than a mile wide, together with numbers of smaller ones, a few of which melted away back from the shore, but which in former times had contributed ice ships to the waiting waters.

The entire front of a small glacier had recently slid into the sea. Its channel was a few feet above sea level. Standing in the rock channel by the broken ice front I could hear the grinding of rocks and ice as the ice slide invisibly forward. Beneath one edge of the front were massed several thousand boulders of assorted sizes. These were grinding against one another and the bottom. At one point, embedded in the ice front, was an angular, unworn rock fifteen or more feet long that had made a long journey without being forced against either the bottom or another rock, though other rocks had been ground to dust under terrific pressure.

Northward, across a narrow arm of the bay, a small glacier up in a hanging valley, the end of which was about one hundred feet above the water, discharged its icebergs with drop and splash into the bay. Hearing a crashing, I looked across in time to see an enormous ice chunk—it was the entire end of this glacier—tumbling into the bay. A gushing, enormous fountain of water shot up and a ponderous wave swept from it across the bay. This wave threw water over the Indian boat menders who were at work more than mile distant and one hundred feet above the shoreline. Near where I was standing, there came a wild rush of waves, logs, and small icebergs. These were flung upon the shore, and many left stranded from one hundred to one hundred and thirty feet above water level. It was the wildest wave that I have ever seen.

It was dark at the end of the second day when I reached camp. The cheerful Indians had fixed the boat and made an excellent paddle. The following morning they set off down the bay, hoping to find supplies and another boat in an Indian camp along the near coast. An inspector would not have given this repaired boat an A1 release, for in rough water it surely would have gone to pieces. Away went the Indians, with two or three broiled fish. I was not allowed to go along, because the craft was dangerously frail even for two. One Indian speeded with the paddle while the other necessarily bailed rapidly, and both were apparently indifferent to the fact that they were playing with death. I planned to remain close to camp, as the Indians felt they would find necessities and return that night.

During the morning, I wandered a few miles southward along the now famous Russell Fiord. It was up this fiord that the Harriman party steamed a few years later. During the afternoon I strolled the shore, watching some of the numerous moving glacial actions. One of the best exhibits of the day was given by a hulk of a flatboat-like iceberg that was top-heavy and tilting with a mass of boulders and other glacial debris. It was dark enough for a collier. It came in sight from behind other bergs, drifting down the bay, with parts of its cargo occasionally dropping over-board. In passing near me, it struck an invisible obstruction and gave a lopsided lurch, dumping most of its cargo into the bay. the dumping of debris, the filling of the bay, was steadily going on.

This berg, an instant after dumping, rolled back and came near to turning a side turtle. Shaking itself as it rolled about, it finally turned end for end. Then this rudderless fresh ice hulk was caught in the outgoing tide and set off for a vanishing voyage somewhere out in the wide salty sea.

Most glaciers over the earth have been shrinking during the last two decades. This shrinkage is due either to lessened snowfall or to a slight warming of the glacier regions. Of all the remaining glacial regions of the world, it is doubtful if any excel the wonderful one round Yakutat Bay.

Glacial debris in inconceivable quantities, with embedded logs, strewed or formed every shore of the bay. One stretch of the shoreline had been recently uplifted by internal earth movements—this was about twenty feet above its former level—while another stretch showed subsidence of several feet. At one place, a grove just submerged was being battered away by the waves.

On the shore, on moraines, and in detached places on

the mountain-sides, were groves of Sitka spruce and growths of arctic willow and alder. I saw many kinds of wild flowers and numerous species of migrating birds. Resident gulls and ptarmigan were plentiful.

During the calm, clear evening I built a bonfire of extravagant proportions. I was determined to give welcome to the Indian rescuers if any returned—the warmest welcome possible for a castaway. As I sat by the fire, I could hear the splash of falling ice cliffs and the never-ending wash and dash of ice-sent waves against shores near and far. Shortly after midnight two boats rowed into the outer edge of my bonfire light.

Three hours later, two boats, four Indians, and I were dodging icebergs down the bay.

One of the large bergs had a number of spruce logs half embedded in it. These thrust from the sides and the top. Flocks of birds rested on these logs. The Indians said that birds sometimes nested on icebergs that floated about in the bay.

We landed on the main coast for the night. While busily engaged in making camp in the edge of a dense, damp spruce forest, a small steamer rounded a forest point about a quarter of a mile down the coast.

After a deal of shouting and signaling, we attracted attention, and in due time I was on board, with the two Indians who took me into the bay and who were to be with me during the summer.

The steamer had brought a number of enthusiastic prospectors and their outfits and put these ashore. Alaskan prospectors were increasing in numbers. Two days later, the two Indians, several hundred pounds of supplies, and I were put ashore at the foot of Chilcoot Pass trail, the trail which became famed a few years later during the strange, intense gold-seeker's rush.

News From a Fossil Lake

A lake has but a transient place on the ever-shifting surface of the earth. A few short-lived lakes are early drained; and very many are slowly buried, or filled with the earth wreckage carried by water and wind. When sifting volcanic ash makes a shift of scene around a lake, fossil treasures take its place. In the ashen shales of the old lake bed are registered many of the details of the event—the what, when, where, and how. By means of these fossils, the life and scene of this lake are restored and illuminated. Volcanic ashen showers in the ancient Miocene epoch filled Lake Florissant and deeply buried the surrounding forested mountains. A fossilized sequoia stump of gigantic size stands as a strange, supreme monument over this primeval Pompeii—over this vast prehistoric ruin.

Years ago, in making a circuit of the upper slopes of Pikes Peak, I came one evening unexpectedly upon this agitated stump. I camped that night by this noble tree in a wilderness of unknown fossils.

This stump would have impressed any one, but there was no hint of the lake's former scenic grandeur, of a wondrous prehistoric ruin; no suggestion of the fossil deposit, a famous geological landmark.

I tramped the nearby surrounding granite mountains, but nowhere could I find vent of volcano that had played the part of Vesuvius to Florissant. But this volcano may have been at Cripple Creek, a dozen miles away. In Miocene times, there was volcanic activity in this region that forced lava into fissures and faults and covered the surface with a gold-filled lava flow.

Trees collapsed beneath the weight of ashes, and their limbs, with leaves and cones, as well as birds from the trees and thousands of insects, were all buried together in the lake and form a telling part of the fossil drama that is read today. Gorged fossil fishes with protruding stomachs show that they were feasting when the ashes ended all. Numbers of these fossils are now on the surface,

"On the slopes of Pikes Peak there is more geology than might be seen in a tour of many a state: sedimentary sandstone, red granite, Ice Age boulders, fossil redwood and fossil reptiles, a fault of magnitude, picturesque erosion and outcrops of iron, mica and other minerals."

Courtesy The Photo-Craft Shop, Colorado Springs.

uncovered and exposed by the erosion of ages. From this array of fossil wreckage, the lake that existed five million years ago is brought again to life.

The fossils show that Fate kindly brought the end abruptly. All colors were flying; the scene was robed in leaf, vine, and flower—stirred with the flit and buzz of life and the glory of summer.

Butterflies fluttered in the air, ants were busy on the slopes, film-winged dragonflies darted and poised, birds sang in the trees around the shore, artistic cattails waved in the breeze along the water's edge, molluscs and fishes lived in the water on the lake's last day, when from the sky the fatal ashes fell.

The region today is of commonplace character, but when Mountain Lake Florissant last rippled in the sun, it was more than five miles long and one mile wide, nearly L-shaped, with ragged, much indented shores. Giant redwood and incense cedar plumed the lake's peninsulas, aspens merrily danced their leaves where the merry brooks came in, mosses, ferns, and cattails decorated the shores of shallow bays; alder and willow may have woven an enameled purple shadow matting around the shore.

The climate was warm and moist. The lake was in the midst of a magnificent forest: giant redwoods—sequoia—and large walnuts; thousands of beeches and three species of birch; a showy exhibit of chestnut and cottonwood; a mingling of vigorous basswood, ironwood, oak, elm, hickory, and pine. Fig and persimmon trees, magnolia and holly gave a luxuriant subtropic touch. Two geologists also report club moss. The smaller forest trees were abundant: gooseberry, sumac, and a wilderness of wax myrtle. Several grasses claimed space, and roses, Virginia creepers, asters, and perhaps thistles bloomed in colored beauty around the primeval shores.

Fossil Lake Florissant is in the Rocky Mountains on the western slope of Pikes Peak, at an altitude of about eight thousand feet. Today, cottonwoods, pines, aspens, willows, roses, numbers of wild flowers, and many other plants growing in its ruins are almost identical with those that grew around the lake. But numbers of species once growing there are not now found nearer than subtropical zones.

Then, as now, everything had enemies to bite it. Caterpillars were present. Telltale holes in leaves of plants, plump parasites, plant lice, pests with bills for sucking blood, poisonous tsetse flies, show in fossil sculpture the life-old struggle for existence.

Other fossils illustrate a better phase of evolution. The forms of many flowers, the apparatus of many associated insects, tell of an old and intimate alliance between plants and insects and show that their lives were interlocked and interdependent. The presence of flowers with petals and perfumes proclaimed to helping insect friends a harvest of honey, the sweetest product of the wildflower world, for the winged love messengers who would carry pollen to the waiting mate to perpetuate and improve the species. In the age-old drama of evolution, in the life customs round this ancient lake, are shown the two contrasting phases of struggle for existence—nature red in tooth and claw, and also cooperation and mutual aid triumphing in a thousand ways.

Here, as at Pompeii, now sleeps the varied life that was. I spent fascinating hours opening unmarked, unlabeled fossil stones, searching for secrets that gave trails or life glimpses of ages past. I opened one day two fossils that had lain side by side for five million years. One contained a matted mass of ants, the other a sprouted seed—a strange case of arrested development. I suppose the seed was bursting when the ashen shower brought the end and cast seed and sprout in stone. The enlivening pulse of spring had throbbed across the earth five million times, but the crystal sprout of bursting seed had felt no stir.

Once I came to this fossil field hoping to find a walnut or other stone seed within the stony shale. I mined with pick and hammer, shovel and brush. I dug up stories, but not the story I wanted. Other diggers found fossil figs and from the flaky shale took delicate and fragile sprays of sequoia fruited with crystal cones. Another excavated a cattail cluster that carried a colony of fungi—it had been preyed upon by pests.

Fossil Redwood stump near Florrissant, Colorado.
Top: Photograph by H.L. Standley.
Bottom: Photographer Unknown.

A fossil is the remains or the result of something that lived long ago—of something extinct. It is a mold, case, petrifaction, track, trail, print, or impression recorded in stone. Most fossils form beneath water level. Mud with tracks change to stone; a log or a carcass is often fossilized by water carrying away the original substance and replacing it with mineral deposit. Sometimes the original matter is carried off and nothing is left in its place; the mold thus resulting shows form and features and sometimes, like sculpture, restores a scene.

For hours at a time, I sat watching a patient geologist as he slowly released the latest fossil news bulletin from the triumphant carrier rocks. These rocks, laden with Miocene messages, had come successfully through the cataclysms, the changes, and the wrecks of years and ages, with news in picture and in cipher. Evolution is the key that unlocks any fossil cypher code. On day, the geologist took from the seal and clasp of stone a broken, full-leafed tree limb. There was also a fossil caterpillar which may have fed upon the leaves. The cataclysm, eons old, had together cast and crystallized the eater and the eaten. I do not know if this be news or tragic nature drama older even than the hills.

Often I searched for a fossil dragonfly. Coming upon a promising lump of brown shale I carried it to camp. Slowly, with scientific care, I chipped, scraped, and brushed off the fossil's seamless covering of time-made shale. Half the form showed a spider. He was embalmed and pickled, stereotyped in stone, but he had the approved form of the present. He was a near duplication of the spiders about camp who then were watching silver nets set for prey.

Erosion—running water—has long been seizing everything for sea sediment, and has carried off masses of fossils. Countless unassorted and unread stones remain. Many of these have been brought to light by erosion cutting channels and removing the surface covering. Unknown numbers having rare value are waiting to be brought from their coverings.

Although this mine of fossil gems has been but little worked, it has furnished the museums of the world more than thirty thousand fossil insects and more than fifteen hundred species of plant, animal, and insect life—a wondrous array of crystal figures set in stone. This fossil display holds, so far as I know, the only butterflies and roses found in the Western Hemisphere. A world's fair of wonders has been rescued from the old lake bed—an exhibit which shows what the toiling ages fashioned and grouped for transient display at Florissant, in the horizon land of Miocene.

It is doubtful if anywhere can be found such a mixed assemblage. Since the days of Florissant, the climate has changed, and plants and animals have adjusted themselves to zones—have become more specialized. Here at one time had lived the sequoia of California, the Tree of Heaven of Asia, the long-leaf pine, the magnolia and the dogwood of the South, together with many kinds of a now far-scattered insect and animal life. The plant life around the old lake was decidedly modern; it shows few changes from then until now; insects show far more change, and since then the mammals show most progressive development.

While fossils were forming at Florissant, they were forming in hundreds of other places over the earth. In a thousand localities, the life and scenes were preserved. Stones were embossed, stamped, printed, and pressed with leaf, flower, insect, and animal—with the records of the prehistoric past. These records, made and deposited in stone, tellingly, graphically give the doings of the place and day. Numberless fossil sources from the Miocene epoch have furnished the facts which make possible the restoration of climate, the conditions, and the life of the world during this epoch.

One of the strangest glimpses brought out by the fossils of Florissant was that there had been tsetse flies at this old lake. The four species found there in fossil evidently were migrants of recent arrival. They are almost identical with the deadly tsetse flies now found in Africa. It is possible that these flies stole a ride and subsistence off the African Rhinoceros emigrants which arrived in America during the Miocene epoch.

The Rocky Mountains then were not quite lifted into

their present place. Forces of erosion were cutting and sculpturing them. But Pikes Peak may be older than the adjacent Rocky Mountains.

The climate of the earth during the Florissant era was warm and moist. The fossils show that there were luxuriant forests much farther north than the farthest present tree growth. Southern Europe was tropical, and there were palms with appropriate animals and other life in northern Europe and walnuts in Connecticut. There were five hundred known species of plants, and no one can say how many unknown ones in America at that time.

The sequoia redwood and the sequoia big trees are today found native nowhere except on the Pacific Coast. But the sequoia redwood had a forest round the old lake, and many of the trees were of gigantic size. There were sequoia forests during Florissant times over the greater portion of the earth—in Canada and Alaska, Greenland and Spitzbergen. It was the common tree of the Northern Hemisphere. It extended southward through Europe and through Asia to the Sea of Japan. It came into existence during early geological times, and was highly developed before the Miocene epoch; and in this epoch it was almost identical with the redwood now living in California.

Often I have visited this land of fossils, where Imagination, greatest of companions, is at his best. Across the fossil lake I have walked alone beneath the Hunter's Moon; by the giant redwood stump I have stood in rain and mist and moonlight, and when its broken column and its accompanying dead were decorated with snow flowers.

On an early camping trip, I found a number of men on the lake's primeval shore trying to cut down the matchless redwood stone stump. This barkless giant tree trunk was fifteen feet in diameter. Judged by existing sequoias, which it closely resembles, it must, when alive, have been not less than three hundred feet high, its age two thousand years or less. A number of the stumps have been chipped and broken and carried completely away by relic hunters. This one has been reduced in height and in diameter by chipping. It is a stump of rare agatized wood, a precious stone of dense and beautiful colored crystals. Thin chips broken from it appear like living butterflies. Just how old the stump is no one knows—not less than five million years. Nature thoroughly compounded its crystals. With all their tools and chemicals the ever-contending weathering forces have assailed it in vain. It is an enduring product of nature's slow compounding laboratory. It is almost as hard and as durable as diamond.

Nor has man cut it down. After days of effort, this stump was left standing, and the stuck steel saw is rusting in its side. It may outlast Pikes Peak.

The plan was to exhibit a segment of this splendid specimen at the Chicago World's Fair. It is worthy of exhibition anywhere—with the best.

Long ago, the annual rings of this ancient giant tree changed to crystal, agate, and opal. It and surrounding fossils reveal the eloquence of ages. It lived a part in a wonderful scene of a wonderful time—in an age which stands out conspicuous in the innumerable periods of the past. Sometime I shall go back again to where the heroic stump has watched the passing ages—where it still watches with all the speechless eloquence of the Sphinx.

Taking the Earth on High

While I was visiting the first settler in the Missouri Valley a spring wind moved parts of the earth and things anchored in it. Spring seeding was just finished and the fields were pulverized and dry. The air was full of dust, and sand was drifting like winter's snow. Real estate transfers were rushing; there was a boom in soil. In places, entire acres of furrow-deep, rich, sandy loam seeded fields were blown out and drifted or scattered over fields to the leeward. When it was all over and things settled down, many a farm had big drifts of sand and soil which plainly had been taken from an adjoining or even a remote farm. Rich drifts beneath boundary fences were seized and were subjects of contention. Floods often go in deep for real estate and hurry away with it, but they generally move along narrow ways. This wind had hit and swept a string of counties. And the interesting thing was that this rich ashen soil that had been moved and scattered was volcanic. As the nearest ancient volcanoes which might have contributed this were a few hundred miles away, the wind centuries before must have deposited this very soil which it now seized and carried away despite the vested titles of many farmers.

As the wind went into high gear, the old settler was reminded of an early day windstorm which he said "tore up the earth, threw things to right and left and into the air, knocked the water out of the river, and blew over trees that never blew over before." The wild behavior of a number of trees in the wind before the window indicated that several might now do something they had never done before. Through the whirling dust screen, the trees appeared in a riot, striking at one another.

The henhouse of my host also flew to leeward, while the chickens, with dignity thrown to the winds, took off on lines of least resistance, frequently wrong side up.

Suddenly the wind went into low and the rain began to fall, perhaps as it had never fallen before. There was a flood, a rush of surface water; there was to the windward, and it was laden with real estate recently transferred by the wind and other associated soil. This large farm was being broken up.

A mass of two hundred heeled-in young apple trees were gullied out and the trees carried through the line fence. A calf pasture fence was not a fixture and it accompanied the calf into the neighbor's orchard. Hallowe'en youths had never wrought such confusion. When the storm ended, my host and his neighbor each possessed real estate and other property with debatable titles. Neither could hold a number of these possessions without the consent of the other, nor recover treasures from across the line fence without the permission of the other. But these two farmers were real neighbors; they had prospered and enjoyed life through cooperation. When I left to take the train, they were merrily debating property rights without "clubs or lawyers," and trying to determine the ownership of certain pieces of property whose titles and transfers had been irregularly and amusingly complicating by a said party—the wind.

The wind is steadily wearing away the surface of the earth. All the land surfaces are regularly swept and often scoured by the wind's steel brush. The wind works alternately with rain and running water in lowering the surface of the earth. As an erosive factor in developing the canyons, hills, and valleys and plains of the earth it is next to running water.

The wind moves all unanchored sand, sediments, and gravel. A sand-laden wind rasps like sandpaper; is a terrific sand blast; it is working with edged tools and is a fluent gouging machine. Winds, thus equipped, make topography, are sculptors, and have helped to shape the natural bridges, the mountain passes, the picturesque rocks, the striking monoliths, and a thousand forms of the statuary of nature that are scattered and grouped over every land. Clear water has no cutting edge, but add a dash of sediment or sand, and its erosive power is greatly increased. So it is with the wind; add dust and sand, and it erodes tree trunks, posts, telephone and telegraph wires, and the earth and rocks themselves.

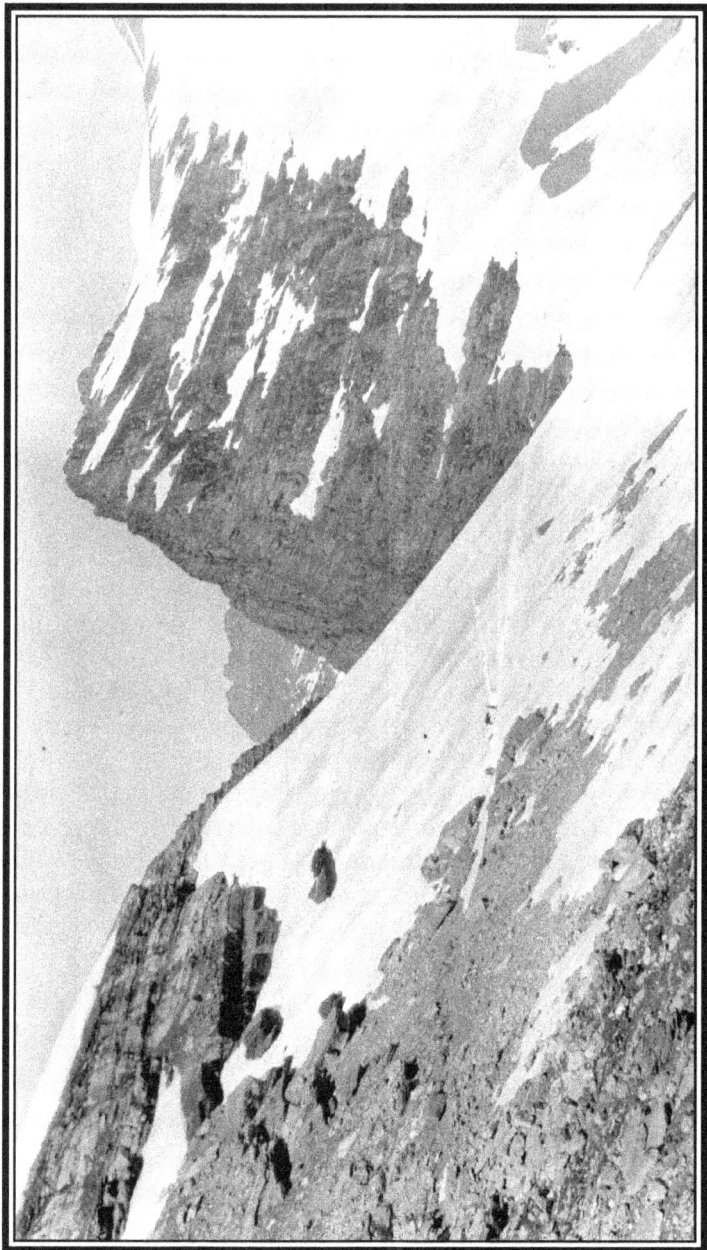

Gunsight Pass, Glacier National Park. In places the wind specializes in pass-making.

The sheer speed of the wind not only loosens and dislodges dust, sand, gravel, and even heavy flakes of stone, but it often transports the lighter material long distances. Tons of dust are sometimes blown hundreds of miles.

Many areas in the United States have been enriched with earthy cargoes carried by tides and currents of air. Wind storms frequently seize the rich and idle Sahara sands, rush these across the sea, over international boundary lines, and then, hundreds of miles distant, drop them as contributions upon the thin-worn soil of Europe.

One segment of China, larger than all France, is deeply, richly overlaid with windblown soil. For ages, the west winds brought this from central Asia. This annual film of yellow dust gave generous fertility to the land of the great moguls. Though for centuries the Yellow River has been carrying this away and building a broad delta into the Yellow Sea, the soil, still rich with west-wind spoils, feeds a crowded population. The Yellow River of China probably receives its name, its color, and much of its sediment from the ancient dust which its waters are washing from this region. After generations of lowering erosion, this dust layer over the Yellow River region still is in places hundreds of feet thick. A vast area must have been eroded for the making of this deep deposit.

A single dust storm is computed to have snatched up and carried off more than two million tons of dust. The Weather Bureau once estimated that during a snow blizzard more than thirteen tons of dust fell to the square mile over an area of more than one hundred thousand square miles. Each snowflake thus was a flying machine carrying a number of earthy passengers. One brief wind storm piled more than a freight train load of dust, trash, and sand upon less than one mile of railroad track—real estate that had soared, and dropped! As I was a passenger on the train held up by this drifted sand, I joined the shovel crew and helped compute the quantity.

During dry winters, the winds which blow across the Rocky Mountains carry trash, dust, and sand. Much of this is deposited in leeward shelters on the upper slopes in

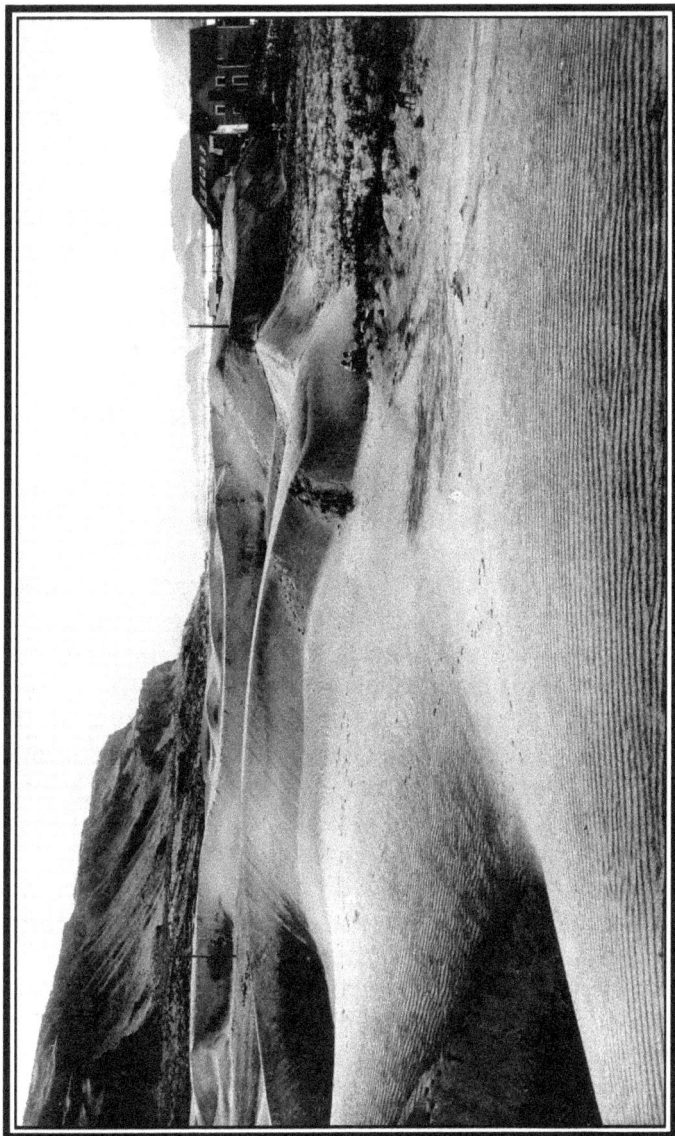

Near Biggs Station, Columbia River, Sherman County, Oregon.
Photographer unknown.

measurable drifts which form exactly as snowdrifts ordinarily form—during the first wind that follows a snow.

I built a little shelter cabin in a windswept opening on a forested mountainside. Returning to this after an absence of several years, I found a windblown excavation by and beneath one corner where the broken winds had swirled hardest. A deep, powderhorn-shaped excavation had been blown out. This opening was seventeen feet long, pointed and shallow to the leeward. The windward end was seven feet wide and about four feet deep. Stones as large as croquet balls had been knocked from this pit by wind gusts.

When the wind blows violently across deserts and other sandy, dust-covered distances, the air becomes so heavily charged with dust that one breathes with difficulty. More than one traveler has been smothered in a desert dust storm.

While in the heart of a Nevada desert I saw a seething, sky-filled dust storm bearing down upon me. A desert dust storm is a scourge, often deadly. I was in the dusty level old lake bottom of Lahontan. One of the precipitous and towering mountains that pierce this lake bed rose high and steeply by my side. I tried to escape the fury of the storm by climbing. The storm front swept wildly forward like a vast prairie fire and filled the western sky from horizon to horizon. Its lower part was an enormous breaker of rolling, splashing sand. Its upper part boiled and rushed high into the heavens like a burning oil tank. With the lower part ahead of the upper, it rushed forward.

After climbing a few hundred feet, I turned to see its front before it swept over. Its base was violently dashing the dark sand into the air, and above this, the lighter dust swarmed thickly. It ripped against the cliff and swirled round me. With handkerchief over my face and half smothered, I climbed. The storm tore below. Not until I was about two thousand feet above the desert floor did I get above the current of the storm, and even at this height, the fine dust filled the air. I was in a moister zone, among pines, and on coming to a trickle of a spring, I lay down to wait for the storm to pass. The moving, dust-stirring pageant was twenty-four hours in passing.

Many a desert caravan has been caught and buried forever by a smothering sand storm. Many a time I have endured one of these withering blasts for hours.

Numbers of fossils have recently been found of animals buried in windblown sand. These were discovered singly and in groups. Four wild hogs were dug out that had huddled together. That several wild horses ages ago sought shelter from a desert storm together in the lee of a butte is the news story told by their grouped fossils. Apparently, much of the West was arid immediately prior to the Ice Age. At any rate, extensive fossil deposits of ancient horses whose original bodies were buried beneath drifted sand have been found all the way from Dakota into north Texas. In places, whole herds have perished.

Desert winds bury with their dust and sand, but often they uncover or erode with their battering sand blasts. On a Nevada desert, I saw a long-buried fossil lake, a salt deposit, having its covering of sand and shale slowly, completely ripped off. The sands of other ages had buried this. Elsewhere in the Nevada desert, I saw a hill of recently piled sand that was more than two hundred feet high and half a mile long.

Increased dryness over vast areas of Central Asia, some thousands of years ago, allowed the desert to extend its holdings. Ellsworth Huntington, Sven Hedin, and other travelers tell of numbers of sand-buried cities, and of some cities long buried now having the overlying sand drifts blown off them. It is well known that, some thousands of years ago, the climate of parts of Asia became drier, and this change compelled a movement among millions of people. The vegetation died and the winds seized and shifted the uncovered, unanchored sand.

The wind is the great seed-sower. The great pine, spruce, and fir forests of the world, the splendid sequoias, and thousands of species of wild flowers all depend upon the wind to scatter their seeds, to help them extend their territory and hold present possessions. In doing this the wind promotes forests and grasslands which cover and protect the earth and defend the surface against wind erosion.

The wind is one of the earth's leading landscape makers and is one of the three great erosive forces and transportation scene shifters. Eroded earthy material is moved by a variety of transportation—slow-going glaciers, running water, ocean currents, wind-propelled waves, and then push and pull of gravity. Most material taken and transported by a glacier is piled in hill-high deposits—moraines; a river fills in and makes low delta plains; while the wind makes sand drifts and dunes, it also scatters these afar, and is more democratic in the distribution in regions where the earth lacks the protective covering of grass or vegetation. Sooner or later, any loosened material will be swooped upon and seized by airplane and scattered everywhere, over lowlands, high peaks, and far out to sea.

The seashore, many lake shores, the broad low-water channels thousands of miles along the Mississippi, Missouri, and other river valleys, the endless sandy distances of numerous deserts—all have a continuous though ever-changing exhibit of the wear and the work of the wind. Many a lake has been formed by windblown sand damming river channels.

In western Wyoming I sought shelter for the night in a trapper's cabin, built fifteen years before. The cabin was steeply tilted in a four-foot hole which the swirling wind had beaten and blown out from beneath one corner. The floor was in place and offered good coasting. Another time, on the lower Mississippi, I sought in vain among the sand dunes for a cabin recently used by levee engineers. I slept under the stars, and in the morning, a brick-red projection in a dune top about a stone's throw distant proved to be a wind-evolved dry-land periscope of the sand-submerged cabin. In revisiting the dunes on the coast of North Carolina I searched for a cabin I had used during a former visit. The wind had remodeled the landscape, or sand-scape, and put a water decoration—a new bay—into the scene where the cabin had stood.

A Utah prospector complained that his cabin during his absence had been covered by a "malicious migrating" dune. He built a second cabin and cut out a few trees so as to divert a possible dune drive. Returning from another prospecting trip, the second cabin was barely visible, but the first one was on exhibition. He moved into the old cabin, and the sand did not again bother him. Although he had an excellent sense of humor, the prospector seemed to consider this cabin burial something of a disgrace. As he said, "To allow such a loose and slow-moving thing to catch a fellow must indicate that he is something of a slow affair himself."

The off-water winds of lakes and seas herd and drive migrating dunes forward, overwhelming groves, meadows, lakes, and streams. Sand dunes are a distinct feature of the seashore of lake and of desert landscapes. In the face of all the objects in the landscape, a sand dune may be considered as most nearly being a moving picture. Their advance may be only a few inches each day. Such advances often bury a forest, and in due time the winds uncover it. Later, this same dune may return with a reversed wind and rebury the dead trunks.

In eastern Oregon, I went out with a pioneer to have a look at a locality where a traveling dune had played an interesting part with a grove which stood on a low plain. This dune originated its materials at a river about half a mile to the east of the grove. Traveling from east to west, the dune advanced toward the grove about one hundred and fifty feet a year. Its front was nearly a quarter of a mile long. This dune finally overwhelmed the grove, but it continued to advance. Slowly, the dead trees of the grove were uncovered, and in due time, the entire grove stood out uncovered and leafless, with a number of the trees limbless from the weight of the sand. The dune traveled on.

Meantime, another dune was advancing from the northeast. I know not just what there was in the shape of the topography that gave this dune its course. When the left wing of this second dune was approaching the dead grove, the prevailing wind appeared to have changed. At any rate, the left wing of the first dune turned about face, so to speak, and started to return over its former course. When I visited the place, it was invading the dead grove from the west side, and a year or two later, the dead grove was again completely buried. The dune was still traveling back toward the river.

Dunes are no respectors of buildings. At Liege, a building has twice been taken down and rebuilt to prevent its being overwhelmed by a broad-fronted dune. Though this building now is two miles inland from where originally placed, the dune is again close enough to whisper, "I'll get you if you don't move out."

A forest fire in Oregon caused the local prevailing wind to change. The former prevailing wind had driven a dune from the river about a quarter of a mile. A wrecked and treeless grove was behind it. With the changed wind, the dune had about-faced and was backtracking through the dead grove toward the river.

Sand dunes of every form and finish adorn the deserts of Nevada. Numbers are of coarse, heavy sand; others of fine texture, in white, brown, or black. Many are as smooth and graceful as an ocean swell and are daintily decorated with wind ripples. I saw a number of dunes in the dry bed of old Lake Lahontan that were going somewhere. The average dune exercises and tumbles about without making any advance. These enormous dunes were in windrows seven ranks deep. They were from thirty to seventy feet high and miles long. Traveling eastward, they were going across plains, ridges, canyons, and mountains, not turning aside for any surface inequality. Eight years later, when I next saw them, they had advanced about two thousand feet. Ultimately, I suppose, they will meet counter winds, be somewhat scattered, but probably deeply bury a region.

Trees on wind-beaten beaches and windswept plateaus have been trimmed, molded, and adjusted by wind-directed sand. On the dry wind-battered heights of the Rocky Mountains, two miles above sea level, the trees have been dwarfed and distorted by these storms. The trees do not surrender, and many are forced as huge crude vines to crawl leeward, and a few that stand lose all limbs and bark on the stormward side and their few surviving limbs stream leeward, making of a tree a tattered, triumphant banner.

The wind occasionally exceeds the speed limit; these

Notchtop Mountain, Rocky Mountain National Park.

may be on the sea or seashore, across mountain plateaus, and also, from the reports of aviators, in the region "between the thunder and the sun."

I climbed Long's Peak on a day when the wind broke speed records. In a pass of 12,000 feet, ir ripped by at 171 miles an hour. The powerful wind breakers whirled and dashed and flung sand and coarse gravel, and several times knocked me down. At Keyhole, altitude 13,000 feet, the wildest place, the wind rushed past. I tried in vain to crawl through it, and finally went around it. I advanced by spurts, between the irresistible rushes of the wind; dropping and holding close until the wave passed; then another dash forward. I rounded the worst point the first try, and dropped on a flat rock as the oncoming wave roared down upon me. This kicked me flying off among the boulders as easily as I might have sent an empty basket.

A high wind often is a natural and terrific sand blast. Where it can seize sharp-edged sand and gravel cutting tools, it blows thee violently against the earth and rocky cliffs. This sharp-edged material, when concentrated by topography against a mountain pass, enables the wind to lower the gradient and to open a thoroughfare. The wind incessantly opposes mountain barriers. It is constantly at work widening and deepening passes across mountain ranges which are effective barrier of easy communication between regions on either side. Both glaciers and streams cut gaps through mountain ranges, but in doing this they work headward and by indirection. On the very summit of a young pass there is little room for either snow or water to accumulate and to work effectively.

In places, the wind specializes in pass-making. The big sky-wide air currents, as they rush forward and come in contact with a mountain range, have their powers converged in pass avenues, and the winds rush across a pass, tearing and sweeping furiously. The loose material is swept out. The removal of this protecting surface cover speeds up chemical weathering or disintegration. Thus through the years the winds have helped make a way—a pass—the skyline highway of wild folks, the passageway for armies, and now the peaceful avenues of mankind. A high wind sometimes is a pass maker—a promoter of civilization.

The wind-driven sand or pebbles, especially in arid climates, wear away the softer parts of the rocky surfaces, leaving the harder strata of the rock exposed in bold, rugged outlines and fantastic shapes.

An example is found in the natural bridges of Utah. Like many other rock forms, they are the product of various erosive forces, but chiefly of wind and water. The material in the bridges being slightly more durable than that of the now vanished rock, or possibly having been less severely tested, has endured while the other material has been dissolved and worn away.

In Little Zion Canyon, Utah, are numerous forms and figures of heroic size and magnificent proportion. In the fashioning of the surface of the earth, Nature in many places makes beautiful, picturesque, and imposing statuary. She has done so here. In the surrounding country are turrets, cisterns, wells, cone-like and dome-like caves and caverns, and nearly complete arches; in fact, arches and bridges showing every degree of completion and past prime condition may be seen in this canyon.

Many times I have seen whirlwinds of magnificent proportions sweeping across the deserts of the Great Basin. Generally they were not funnel-shaped, but ever magnificent columns and large enough and high enough to support the sky. One of them moved strangely across the desert floor one day without a sound from it reaching me. It appeared, and probably was, a hollow column inside the merry-go-round of dust not less than two thousand feet high and I believe three hundred in diameter. For a time it whirled forward perfectly vertical; then it tilted forward, then backward, and finally bent like a pipe about halfway up, and the upper part leaned to the right while the bottom half was vertical.

Another day, while up three or four hundred feet on a mountainside examining an ancient shoreline, a gigantic hollow-column whirlwind burst from behind a high mountain. It was thick with sand and powdered alkaline dust. As it swept before me, another, still larger, hove in sight in the southwest as though trying to interfere with the first one. A few seconds later two others dusted in from the southern quarter and came swirling into the field of action. The first and the second cut an "X" crossing, the second crossing the line as the first cleared it. Their tops were thrown into slight confusion by conflicting currents, but each quickly reformed and went indifferently on its way.

As these four were disappearing, a fifth towering one rolled in, closely following the track of the first one. This passed close to me, and from my height, about four hundred feet above its base, I had a good look at it in action. Twice, for a short distance, it developed a funnel point at the bottom, feeling right and left with this, like an elephant trunk; twice its top was explosively wrecked from invisible, unknown causes. It made an impressive show in the sunlight, and I estimated it to be not less than three thousand feet high.

Ofttimes fine dust or volcanic ash is long suspended or is long afloat and adrift in the air before reaching the earth. Volcanic ash is said to have been carried round the world, and I have seen desert dust in the air nearly five hundred miles from the point of take-off. Sometimes, after a prolonged and violent desert dust storm, the air is hazy, does not entirely clear for days. Bits of thin flaked mica, ash, and finely powdered rock flour are seemingly lighter than air, and linger in the air as though in solution.

Once blown from the earth, every dust atom becomes homeless—a sky tramp. It may wander the sky for weeks. Ascending currents carry it up; another carries it into the north; it is carried back, it drifts and finally settles to the earth or is carried dashing down in a raindrop or down lightly as an adventurer on a snowflake. On earth it dries and wanders on—it is blown hither and yon, there is no resting place for it.

Dust settles everywhere over the globe. It discolors the white mountaintop snows, dusts inland forests, reaches ships in midocean, and is found at rest in layers over the ocean's floor.

The magic sheets and hills of running sand are born of the wind and of the sea. They are wandering, migrating, homeless—restless as the sea. These sands are from lands

"The Crest of the Continent" by H. J. Cowling.

uncounted ages old; they have had a place in many a geological stratum, and will have a part on geological horizons yet unformed.

Birds and animals migrate, emigrate, or travel in far circles and return. Sand dunes, companions of the international, universal winds, may sometime find a resting place—a permanent mound beneath the grass, the lilies, and the oaks.

But many sand dunes are the changing playhouses of the inconstant wind. They are just happy travelers, merry russet-brown gypsies; they never had a home—they do not want a home—they do not remember where they came from—they are not going anywhere—they are not worrying about arriving—in No Man's Land they are traveling merrily and gracefully.

Interviewing Geological Characters

By an ancient water hole in the mountains of western Nevada, a set of tracks showed that two big beasts had clinched and fought! Backward and forward, to right and to left, they had struggled, making huge dots and dashes in the mud. The beast forcing had made tracks set deeply at the toes. The fellow opposed had sunk his feet deeply at the heels. Each had sat back in the mud, leaving there his imprint. And a big rough smear was impressed with marks of a coarse hair coat where one had fallen or been thrown. These tracks were made in mud, now stone.

Spaces around the water hole are cut and confused with many kinds of crowding tracks. There is the hoofprint of a deer and the padded track of a wolf, and in the mud the ponderous mammoth stamped his footmark large. Here this pictured story in stone ends abruptly.

This solid rock mold shows the surface of a deserted and ruined shore of a prehistoric water hole. It carries the footprint records of animal visitors that long ago became extinct. These wild life signs, made a million or more years ago, are almost as plain as tracks in recent snow. Lodged against each squeezed-up track ridge is a dash or pile of windblown sand. An expert geological trailer and tracker said of this imprinted topography that, at the time, or shortly after the last animals were at the water hole, the wind blew strongly from the southeast, and perhaps from a desert.

The fossil of an Alaskan mammoth and the broken bones of wolves in the debris at the foot of a cliff indicate that ages ago, while the mammoth and the wolves were fighting on the bluff, the rim caved off and carried them into one burial place at the bottom.

The wondrous story of Geology, printed and illustrated with fossils in the many-colored rocks, carries enriching treasures for the imagination. This story makes all the

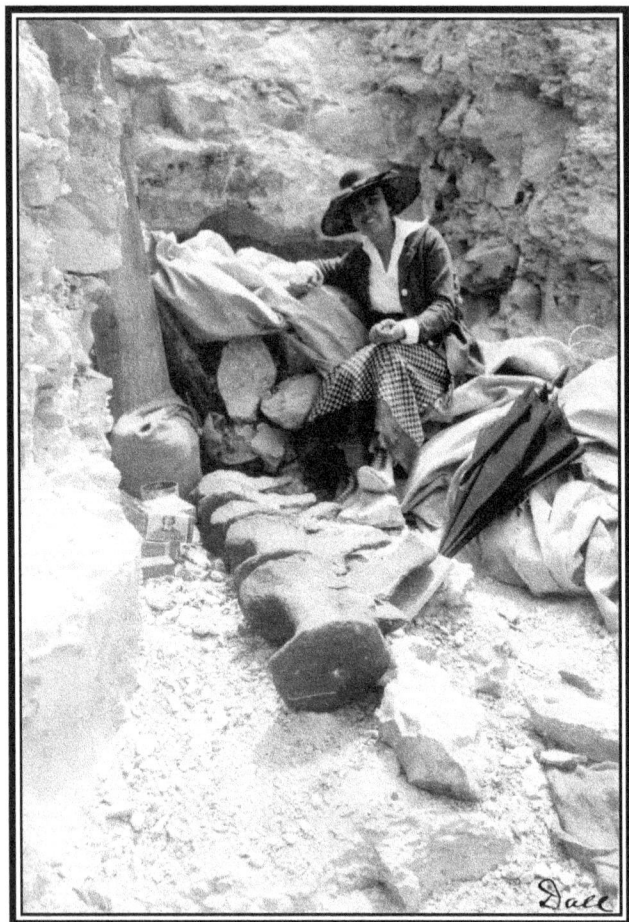

"Part of the skeleton of a Dinosaurus I am
uncovering in Fremont County, Colorado.
Yours Truly, Dall DeWeese, 1916."

world a stage and has a story for everyone.

I never ceased to wonder about palm trees and tropic animals having once lived contentedly in western Kansas. That more fossil remains of prehistoric grizzlies than bones of existing ones have been discovered, and that the grizzly had come over a long trail from Asia into America about a million years ago, was always of interest when I was trailing this wide-awake and adventurous fellow.

In all my camping trips over the continent, I never failed to turn aside for a locality said to contain fossils, and occasionally I had a week or more with fossil hunters. I was out with a fossil hunter in Wyoming who restored the past so tellingly that, although we were working with fossils of alligators and palms in dust and alkali, the moist, warm ages came back and the land of bayous with reedy islands again was seen.

In Oregon, on encountering an outfit of fossil hunters out for prehistoric horses, I gave up an adventurous drive with cowboys after horse thieves. Here, one day, the fossil digger uncovered a tiny horse across which lay a heavy fossil tree. Crushed beneath both were found the fossils of a number of plants on which the little horse may have been feeding. A tragedy of the wilds that happened one million, five hundred thousand years ago!

Small horse-like animals occur in the early fossil records of both continents. Horses appear to have originated in America. Ample fossils here show Eohippus as a tiny horse, a foot high, countless ages ago, and in this continent the horse lived and evolved through many changing ages, numbering many species. A Texas species was larger than the heaviest modern draft horse. Horses numbering millions, as highly developed as the horse of today, vanished suddenly from North America with the coming of the Ice Age. It is possible that climatic changes were particularly favorable to the rapid multiplication of the tsetse fly or other pests. Extensive areas in South America and Africa today are rendered uninhabitable for horses by the tsetse fly, tick, or other plague.

At the time of Eohippus, western North America was a country of many comparatively shallow lakes with marshy flats and rank vegetation. The climate was warmer than it is today. Little Eohippus was one of a host of animals, some of them quite different from present day animals. He was insignificant compared with his bulky competitors, and it is surprising to find that his descendants have survived through the ages, while many of his contemporaries have become extinct.

When the moist country of the West passed through changes and the marshy vegetation was replaced by dry uplands with hard grasses, the horse adapted himself to new conditions. Like every other life that has escaped extermination, the horse practiced evolution, and through the relay race, ages long, kept only the best received from its ancestors. His teeth became longer and enamel-covered, and were more effective for rapid grinding. In the demand for greater speed in the longer distances which he must travel for food and water, he placed more and more of his weight on his middle toe, and the number of his toes was finally reduced from five to one. It took three million years for the horse to pass through these stages of evolution. Few animals have so ancient a family tree. With all of these gradual and progressive changes, there has been a harmonious coordination which fitted the horse to meet competition and to respond to changing conditions of food and of climate. Eohippus was probably as exquisitely adapted to the conditions of his time as is the horse of today to his environment.

In the development of animals, the great interest centers around the teeth, the question of making a living. The porpoise has opportunity for toothache with 256 teeth, but most of the ancient animals had fewer than 50, and the tendency of time has been to reduce the number. Nature appears to have spent enormous time and skill on teeth models. Not only was the size of the teeth greatly increased as their number was reduced, but the grinding surface, both in area and in powers, was markedly increased. This has been of inconceivable advantage to the horse. Nearly every species of animal that has avoided extermination has shown a steady change and improvement in its teeth from age to age. Birds have, of course, lost their teeth, but they

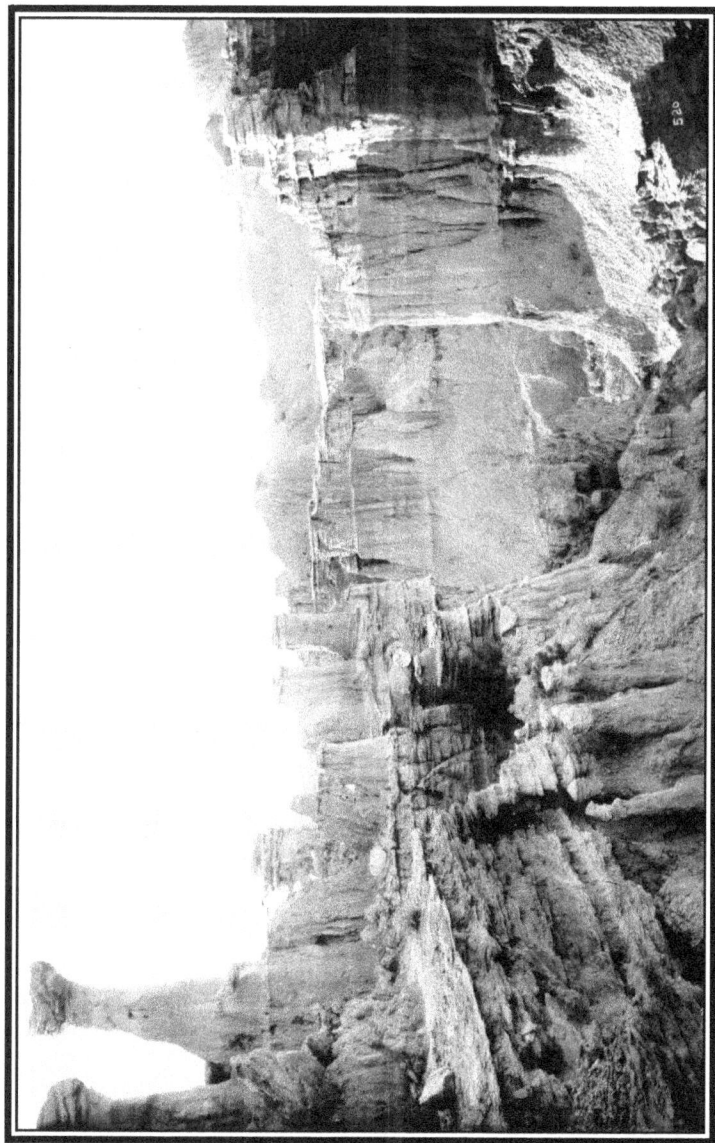

Protoceras sandstone area, Big Bad Lands, South Dakota.

changed their habits, developed flight and use of air, so that in their case the loss of teeth was an advantage.

The teeth are not only an excellent key to the food habit of the animal, but this in turn has a bearing on the bone arrangement. The structure, jointing, and mechanical plan of most skeletons fit the possessor for his peculiar food-getting habits. This would tell where he ranged, in swamp, plain, or forest, and his environment would generally determine his color. Animals of the plains have unbroken, neutral color; forest animals are striped and spotted; desert animals are sandy brown; those of snowy regions, white. Usually, color conceals or renders the possessor inconspicuous, but sometimes it serves to reveal his presence. The tail feathers of certain birds and the tail of the cottontail rabbit are white probably to reveal the individual to others of the species who may be following.

No life on earth, neither the old sequoia nor the whales—the swimming mammals of the sea—so arouses the imagination and gives it romance and vision as the story which fossil facts picture and restore. Paleontology reveals the zoology of the past—one hundred horizons once filled with life, movement, and song. Evolution, through the slow eons, has brought into being an array of forms as numerous and as uncountable as the stars. All present forms of life are a story of long, prehistoric, progressive development from simple to the more complex forms. In the process of evolution one species frequently split or divided into several species. Many a species has multiplied and evolved so rapidly that, by the information thus far available, there is no accounting for its swift progress, and, on the other hand, species have been exterminated rapidly by factors or conditions not now known.

The earth was lifeless for inconceivable eras. Then dull and lowly life existed in the sea almost infinitely long before it ventured forth on land. This life evolved; the age of fishes came, and, after epochs, the age of reptiles. During the latter part of the Reptilian Age the animals living in spiral shells developed shells of the most remarkable complexity, diversity of form, and ornamentation. But of all the great hordes of these shelled animals of this period, only one simple form survived to the next.

"One of the most dramatic movements in the life history of the world is the extinction of the Reptilian dynasty which occurred with apparent suddenness," says Dr. Henry Fairfield Osborne. Whatever the cause, the appearance of flowering plants coincides with the disappearance of the high dinosaurs of the Reptilian Age and with the oncoming of the Age of Mammals.

And thus life went continuously on. Meantime, somewhere on the earth, often over a large area, many kinds of life that could not readjust closed their career in a vast tragedy—many species became extinct. Giant salamanders dominated the swamps and bayous of southern Texas some million years ago. These prehistoric fellows, equipped with both gills and lungs, made a big showing, but they had their day.

The constant change and improvement in life forms probably is due to competition, to climatic changes, and to the ability to readjust most quickly to unfriendly conditions or to take advantage of new, rich opportunities. Ages ago, creatures existed that were a combination of snakes and lizards and had the power of flight. These are now extinct, but we have more highly developed snakes, more highly developed reptiles, and more highly specialized winged folks. The chalicotheroid is now extinct, but Time appears to have taken the chalicotheroid claws for another animal, his body for the horse, his head and neck for the giraffe. While he, as a species, has become extinct, his striking special parts have been highly developed through specialization.

Evidently, many species of animals originated in a segregated section, and there remained and evolved for ages. Then that portion of the continent either sank, or the fossil records of those animals have not yet been found—a fossil realm which geology has not yet discovered.

Many fossils were destroyed through excessive internal heat, others were lost through erosion of the rocks which carried them. Other fossils still are deeply buried, and it is in every way probable that through new fossil discoveries

Wind-sculptured Jurassic sandstone. Seven miles
south of Hot Spring, Black Hills, South Dakota.
Photograph by Darton, circa 1900.

the forthcoming chapters of the amazing story of geology will be far more varied and eloquent than the present.

It was by mere chance that fossils were formed, and the many records of ages thus deposited and perpetuated. Fossils have been formed in bogs, water holes, caves, earth fissures, quicksands, landslide debris, and beneath deposits of glacier debris, windblown dust, or volcanic ash. Numerous fossils have been found in the sedimentary and delta deposits of freshwater lakes, but the overwhelming majority have been found in deltas formed by rivers that deposited fine sediment in the edge of the sea.

An elephant is lost in the mire, a flood drowns and sinks an ancient horse in a delta. An antelope settles into the quicksand. A landslide buries a deer. Sand in a desert storm drifts over and smothers a camel. Numberless birds, insects, and animals fall into a crevasse. In each case, the animal is buried in a short time. The remains are slowly decomposed and removed and the spaces they occupy filled in with mineral deposits, and a mold of the original form remains.

Seldom is the original substance of the prehistoric animal as well preserved as in the case of the mammoths of Siberia, the flesh of which, frozen in ice, has been eaten by dogs and wolves. Usually, the original substance of the plant or animal has been carried away in solution by underground water, leaving a cavity in which only the external form is preserved, or has been replaced by mineral or some other material. Nature has made many fossil casts by filling with stone or mineral the cavity once occupied by some form of life.

A fossil may be the solidified remains of some plant or animal. In the fossil beds of the John Day region, a large lizard mother, evidently injured by the assaults of a monster, was found at bay with her baby beneath her arm. The fossil monster lying by them may have caused this lowly mother to die for her young. These animals were buried by volcanic ashen showers.

Thousands of fossil bones, representing many kinds of comparatively recent animals and birds, are found in an asphaltic deposit near Los Angeles. Exuding petroleum evaporated and left a tar that was extremely sticky. This trapped large life as effectively as sticky flypaper holds small insects. In one corner of this deposit, an elephant, a sabre-toothed tiger, a wolf, and a carnivorous bird were found together. Evidently, the tiger and the wolf had come to feast upon the entangled elephant and became skeletons where they banqueted. Likewise the bird. A majority of these skeletons are of animals young, old, or diseased. The young and inexperienced were easily caught, and the old and disabled lacked strength and endurance to escape.

A million or so years ago, before the beginning of the great Ice Age, the earth was probably in the golden age of plant and animal life. During this epoch, bulk appears to have begun to decline and agility and brains measurably to develop. Geologists speak of the fullness of mammal life during its day. Red-blooded animals were everywhere winning. The Miocene epoch ushered in the modern plants, the grasses, the horse, and the splendid age of mammals. Evolution was just finishing many modern forms and was beginning to shape numbers of models that now have a place in the sun with us.

Many kinds of animals which we now have in America were in existence here at that time, and numerous other kinds which we do not now possess. Giant beaver fed on the bark of aspens and cottonwoods and made their picturesque homes as now. Horses, elephants, and wild cattle grazed; bears and lions prowled and growled and roared. The American mastodon was abundant in the forested regions, where he did not become extinct until a very late period and probably after the appearance of the early Indians. The Siberian mammoth migrated by way of Alaska and was abundant all across the northern United States to the Atlantic coast. The Columbian elephant, which attained a height of eleven feet and rivaled the largest African elephants of today, ranged over all the United States, including Florida and even the tableland of Mexico. The huge Imperial elephant, which was more than thirteen feet in height, was especially adapted to life on the open plains and roamed from the Pacific Coast to the Mississippi. Tapirs were not uncommon in the forested parts of eastern North

Badlands just north of Flour Trail, Big Badlands, South Dakota. Photograph by Darton, 1898.

Eroded Forms. Big Badlands, South Dakota.

America as far north as Pennsylvania. Ground sloths migrated from South America and were very abundant in the forests east of the Mississippi. They were great, unwieldy, herbivorous creatures covered with long hair. One of the latest of these to survive was discovered and named by Thomas Jefferson.

Birds were numerous, and hundreds of species then in existence are with us today without noticeable change—among them, soaring eagles, plover, quail, and waterfowl, as well as birds with larger, longer limbs than horses, and head held higher than that of the proudest mortal, which produced eggs, a single one of which might have been sold by the gallon, each egg carrying the raw material for six dozen omelettes. South America about that time had a seven-foot wingless bird, now extinct.

There were mastodons that could consume a grove or a haystack at a meal; and ancient lumbering lizards two and three rods long, which, if they now lived and sat down dog-like, could look in at the third story window. Much of the past is dominated by a procession of giants, both on land and in the sea.

Many of the giants were fellows large of bulk and small of brain. These dull and ponderous fellows, though unarmored, pushed their weighty way down long ages, but finally fashion put them into the scrap heap. These huge and hideous armored fellows must have made a show! Many feet high, many yards long, heavily armored, their sheer bulk should have rated them in tonnage. Low-geared, bristling with plates and horns, they were the tanks of their time. But with the coming of higher-geared and more flexible models, especially mammals—those of hoof, who had better feet and brighter brains—these heavy, awkward, witless fellows, of whom there were great numbers and many models, appear to have gone out of existence as abruptly as did the passenger pigeon. Their slowness of movement, together with tons of food required, must have meant starvation for many whenever food was scarce. All small-brained giants perished long ago. Professor O. C. Marsh proposed the following for the tombstone of one of the dinosaurs—"I and my race perished of

Brule Clays, Badlands north of Scott's Bluff, Nebraska.

overspecialization." The inability to readjust to changed conditions is frequently the price of existence.

Whales which spent their early existence on land may have saved themselves by going down to the sea and mastering water transportation and adopting a new food supply.

During past ages, the earth was in many respects as it is now. There were mountains, valleys, canyons, lakes, and plains. There were forest and willow swamps, open prairies and dry plateaus. In these scenes, animals romped and grazed, fought and lived and died, just as now. Each age had its climate and held upon its horizon an array of wild birds and animals together with innumerable kinds of plants and trees. There came climatic changes that exterminated some species and slowly reduced others, but numbers in somewhat altered and improved form increased in the next succeeding age. The number of species materially increased down through the ages, owing to the splitting up and branching of plant and animal families.

One of the broken, fragmentary, and startling fossil news stories of the middle Miocene epoch tells of dissatisfied folk called Primates. They abandon treetop life, get down upon the earth, begin walking, become the original progressives, project themselves as radiating explorers, migrate and emigrate, evolve. They develop, take on form and feeling, and become man, whose triumphs are yet to be recorded in the higher and more glorious horizons.

At that time, the climate was similar to what it is today. The earth was richly clad in vegetation. There were swarms of insects, butterflies with jeweled wings, acres of brilliantly colored flowers, and many kinds of birds that sang in the trees and flashed their colors in the sunlight. Prairies and sunny meadows were crowded with luxuriant grasses. Forests, vast and varied, contained nearly all the species which we have today and many other species now extinct. There were numbers of elm, linden, cedar, hickory, sycamore, walnut, grape, alder, birch, willow, redwood, banana, fig, magnolia, palm, yew, maple, cottonwood, persimmon, dogwood, holly, breadfruit, pines, and spruces.

The story of prehistoric life is one covering enormous periods of time and is a mingling of tragedy and triumph. Climatic changes, unequal development in adjacent districts, enforced migrations, produced endless clash, competition, war, and struggle for existence. It seems as though everywhere nature was red in tooth and claw.

In fossil bones, many of them large, I have seen those that were ankylosed—had been broken and healed again. So down through the ages, as now, there were accidents and fights, broken bones and suffering.

Then there were storms, floods, and tragic earthquakes that claimed millions. There were appalling fires. There were volcanoes that dealt far-reaching death. But the warfare for existence was not all deadly, not all fear and tragedy. Animals appear to have found repose and to have hastened their development through play. There was mutual aid and cooperation. Might did not always win. Powerful tyrants sometimes lost. There were sacrifices that saved. Every horizon of prehistoric life had its horrors, and countless are the unwritten stories of stirring deeds that must have helped develop life and the growing mind of evolving man.

Environments show foremost in the structure of the rocks holding the fossil. Fossil reports show that the climate of the whole world grew cooler and dryer during the Miocene epoch. This increasing dryness and changes in environment brought changes in the animal world.

With a change in environment and climatic conditions, numerous species of animals emigrated, went on long journeys, and found homes on other continents. Europe and America, about four million years ago, were suddenly invaded by true mammals. This invasion is a geological surprise. No fossil forerunners had indicated their coming —told their presence anywhere on earth. Well-developed, vigorous, and confident, they rushed upon the continent like cattle long corralled. Where were they from? Probably from an old and extensive area on what now is the Continental Shelf between Europe and now is the Continental Shelf between Europe and America. This has

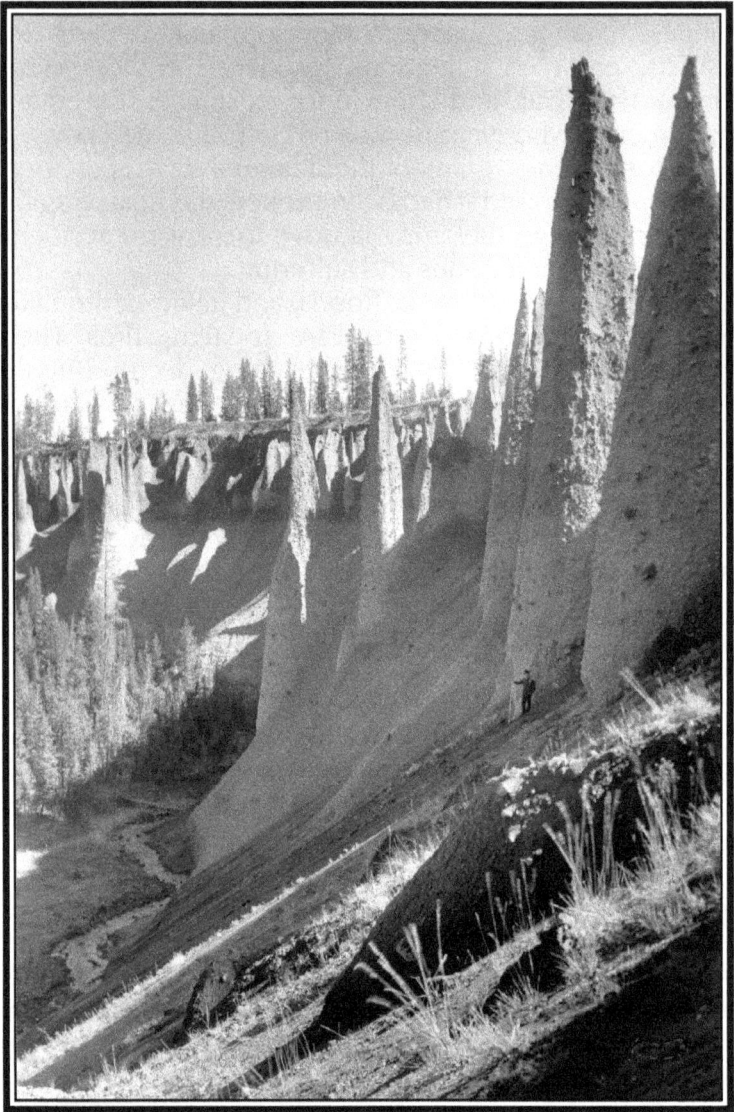

Pinnacles, Sand Creek, Crater Lake National Park.
Photograph by Fred H. Kiser, Portland, Oregon.

only a shallow covering of water, and this, or parts of it, have been above sea level many times and united Asia and northern North America, and North America and Europe.

There is a strange story which follows the forming of land bridges uniting continents. Possibly many species of animals and countless numbers watched restlessly along the shores for these emerging bridges to be completed. Eagerly they may have rushed across; possibly one by one, in groups, in flocks, and in herds. They may have crossed in a mixed multitude, jostling one another. They may have hesitated and retreated. But it is certain that numerous species crossed over.

The entire camel family was confined to North America for perhaps three million years. It appears to have migrated to the Old World and into South America. Mastodons and rhinoceroses and probably many other mammals migrated to America from the Old World by means of the Alaskan land connection. With the joining of North and South America, horses, llamas, deer, tapirs, and others migrated to South America, and the giant sloths and other South American forms moved north.

In crossing these animals may have met other animals from the opposite shore who were complete strangers to them, and who, like themselves, were restless for new scenes or were eagerly searching for a more abundant food supply. The migration of species into new territory has had a tremendous effect on evolutionary processes. It meant increased competition for food, and more alertness was required. Many immigrants failed to establish themselves and became extinct.

There may have been chaos and consternation when the bridges were broken or sank from sight. With the closing of the great highways across the intercontinental bridges, the interchange of wild life inhabitants ended, and the life on each side of this new water barrier commenced to evolve under different conditions.

It was a cosmopolitan world in the early Pleistocene period, with immigrants from the Old World and from South America mingling with those species which were native to North America. Then, as now, rugged mountain

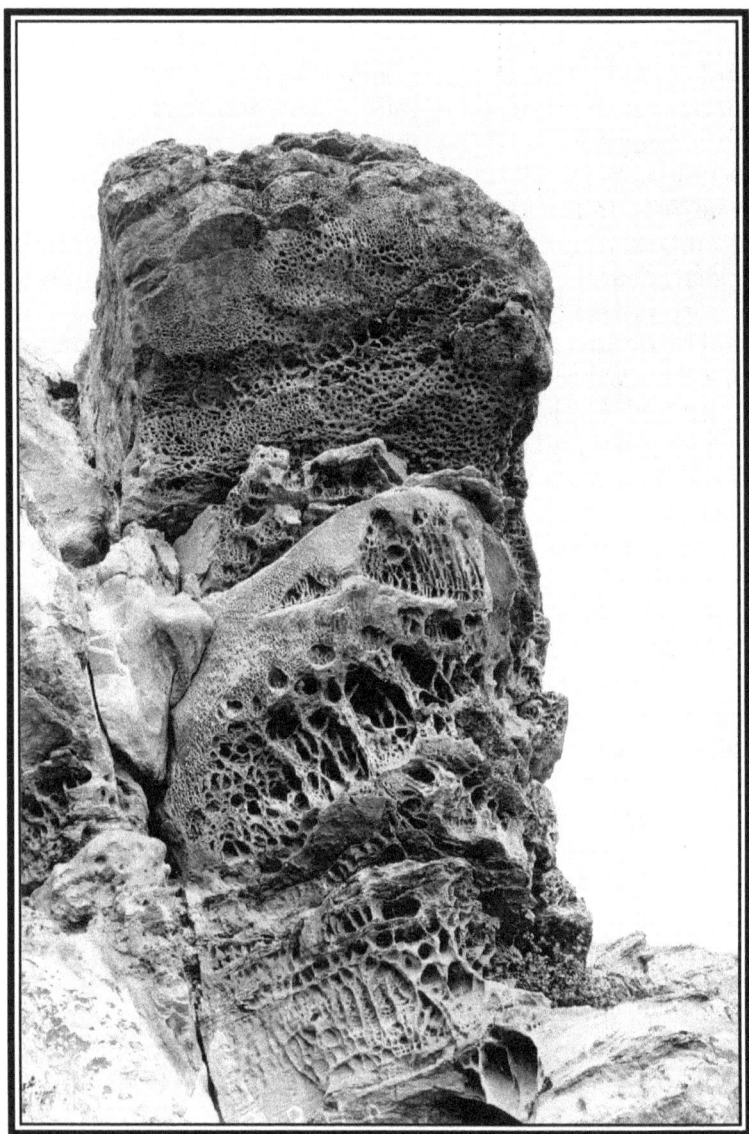

Eroded Carboniferous Sandstone, three miles south of Livingstone, west side of Yellowstone Rise, Montana.

ranges held their snows; the streams flowed on, carrying sediment to the ocean. The great salty sea, with waves and tide, washed the shores of strange continents. In the sunshine, clouds filled the sky, and the rainbow glorified the fading storm. The winds roared, and the mysterious and magic seasons came and went and rolled on as wondrously as now. Somewhere through the scenes at that time, possibly in numerous localities, primitive man was heroically meeting conditions and fighting his way upward.

The period in which man began to develop as a species is not known, but this must have been ages ago. Geologists are not agreed concerning the exact time that should be named for his earliest records. Some would place these back nearly a million years. Dr. Henry Fairfield Osborne says that there are positive human records for 125,000 years. Twenty-five thousand years ago, there came into Europe men much like the classic form of historic Greeks.

The varied and extensive remains of fossil animals that flourished on the continent before the last Ice Age shows that there occurred an extensive extinction of animals which probably was caused by the conditions of the glacial period. Sometimes the animals could not endure the slight decrease in temperature. Often, the food supply was diminished by this change in temperature. Climatic change in old territory often is unfavorable to a few species of plants—and to the few species of animals that feed upon these plants, and possibly to the beast that prey upon these animals. Temperature had a marked effect on plant life and, in turn, upon animal life. The animals which survived generally met the demands for better brains and better teeth and sometimes for increased speed. The remarkable success of the horse in life history was largely due to the development of better teeth and to increased speed.

The Ice Age was a revolutionary and evolutionary factor of first magnitude. All the northern part of North America was covered with ice. This may have exterminated countless numbers of animals and forced millions of others down among those already living in perhaps the crowded southern half of the continent. This made an overcrowding and confusion together with a marked, trying change of

Norman E. Smith, brother-in-law of Enos A. Mills, with a basalt ventifact in Death Valley, California.

Toadstool Park, near Adelia, Nebraska.

climate and environment.

Among the animals species which perished were two large species of the cat family; four species of bears; two species of the sea cow; six species of the horse; the existing South American tapir; wild dogs; a species of llama; a camel; two species of bison; three species of sheep; two species of elephants and two of mastodons; a huge sloth as large as a rhinoceros.

The horses, camels, and elephants which lived in North America before the Glacial period were found afterward only in the Old World. America, ages ago, was richer in wild life than at the present time.

The sedimentary rocks with their fossil figures and plaster casts of perished forms have guarded their exclusive game so well that only now and then has a wandering poacher procured a prize. Recently the fossil hunters, the mightiest the world has ever known, have gone triumphantly forth, and now an open season is declared on all the known preserves and places of prehistoric life. These paleontological hunters have brought in the big game from former seas and plains. They have tracked the horse through many ancestral times; across changing topography, strange horizons, and in different climates, they have followed him for a million years and a day. They have secured homegrown elephants, none stouter or larger; camels and tigers, together with their antique neighbors, the primitive plants and other natural resources from the scene, and an abstract of the annual climatic conditions in which they grew.

Big as life, these prehistoric animals stand in museums, exhibiting their customs, taking again the part they played amid their ancient scenes. These splendid groups and heroic figures set the creative mind to work, and imagination is stirred to explore the scenes of a thousand wonders. Fossil hunters have brought museums to life, they make the past a servant of the present, and awaken to action all the senses that supply rare and rich material for the adventurous brain.

The Oscillating Seashore

An old channel of the Hudson River has been traced by numerous soundings for one hundred and twenty-five miles off the New York coast. This submerged channel extends beneath the shallow waters of the sea from the present mouth of the river in a southeasterly direction to the edge of the Continental Shelf.

New York Harbour is chiefly the result of slight submergence which took place perhaps at the time of the sinking of the channel of the Hudson. Uplift this area slightly, and a number of islands now invisible, and much more of those slightly visible, would come into view and restore geographic conditions prior to submergence. The St. Lawrence, the Delaware, and numbers of other rivers have had their ocean connections shortened by the sinking of the Continental Shelf.

Just what caused this slight submergence along the Atlantic Coast line is not known. A number of geologists hold that the accumulation of enormous weight at the shoreline—deltas at the mouths of rivers, and moraines piled by glaciers—may depress the underlying sea. At any rate, the morainal matter in Long Island, transferred by glaciers of the last Ice Age, is of inconceivable weight.

A slight uplift of the present coast opposite New York could expose this old river channel and a wide stretch of the Continental Shelf. An uplift of less than one hundred feet would cause Father Knickerbocker to make a long journey by land before he could go down to the sea in a ship.

The shifting ocean line at Atlantic City has caused the hotel owners no end of trouble and expense in trying to keep their hotels moderately near the waves.

The Norwegian peninsula has measurably uplifted itself in the last three centuries. Many, and perhaps most, localities of the earth have been uplifted or, through settling, have sunk below their former level. A number have been both upraised and lowered, and a few have had numerous ups and downs. The finding of sea shells in the rocks in Iowa shows that an ocean once covered interior North America. In north Greenland, very fresh shells are found in shore sand one hundred to two hundred feet above sea level.

The geological rock layers upon which Paris stands show that the region has been raised and lowered at least ten times. Each time that this was submerged below sea level, sediments were formed upon it; each time it was above sea level it was carved and lowered by erosion. The relative ages of fossils found in its sediments, and the character of the overlying rocks, enable the long history of the region to be read.

Near Naples, Italy, are the ruins of an old temple which are known to have been above sea level in the third century. Then, for many centuries, a subsidence of the shoreline carried the foundations down twenty-one feet below sea level. The submerged portion of the column was bored full of holes by marine animals. These ruins now stand above sea level—the shoreline was again uplifted.

Near San Francisco, the surface is thought to have ranged from 1,800 feet below its present level to 400 feet above. Boats enter Golden Gate into a submerged river valley.

Along the shores of the continents, shallow waters rest upon a submerged coastal plain. The Continental Shelf or Plain is from on to more than one hundred miles wide. Over it, the water is shallow, everywhere less that one hundred fathoms, and over the greater area, less than one half this depth. Beyond the outer edge of the Continental Shelf the ocean basin drops off to great depths.

This Continental Shelf is sometimes called a submarine delta, for it is a delta. It is formed of the material pushed into the sea by rivers and glaciers, and blown into it from inland or broken from the seashore, as this is invaded by the sea. All these materials are mingled and distributed more or less evenly along the sea coast. This shelf in reality is an enormous fill in the edge of the ocean, rising almost to the surface of the water.

Many forces and factors determine the shoreline of the

sea. Glaciers have in places built enormous, blunt peninsulas out into them. In numberless places, too, rivers by filling in have pushed long stretches of shoreline out into the sea. The sediment which the Mississippi carries to the gulf averages more than one million tons per day. Wind and the sea sometimes build the land outward into the water and sometimes they tear the land away and allow the sea to extend inland. The shoreline of the sea has in many places repeatedly changed and even now is slowly changing. In places, the land at the shoreline is rising, and in other places it is sinking. Los Angeles boasts seven historic beach lines.

The filling in of the oceans with volcanic piles, of which the Hawaiian Islands are an excellent illustration, must tend to cause the water along the shorelines to rise. The wash of vast quantities of sediment into the sea must also fill in the sea basins and force a rise of its waters.

However, all these vast deltas in the sea combined in one would make a comparatively small fill, of such vast size is the basin of the seven seas. In fact, if all the land now above the surface of the sea was dropped into the water evenly over the ocean floor, the average height of the ocean would be raised only about seven hundred feet.

Rivers with vast deltas have filled the ocean's edges and added empires of land surface. The Ganges River, building out a delta 50,000 square miles in area, has loaded down the ocean with thousands of tons of rocks. Somewhere this is bringing a strain on the earth's crust. There is a change of equilibrium.

There have been uplifts that caused local waters to retreat; and volcanoes have given birth to islands—lost tiny continental children in the sea; but, on the whole, the gains of either sea or land, though most picturesque, have been more of the nature of small temporary raids.

The total area of the earth's slightly submerged Continental Shelf is vast, perhaps ten million square miles. An upheaval of only a few feet over an extensive area add greatly to the land areas of the Continent. America and Europe, and Asia and America have been repeatedly united by slight upheavals of the Continental Shelf and then broken apart by slight subsidences. The islands of Bering Sea have risen again and again, forming a bridge between Asia and Alaska, and later subsided.

The draining off of the water of the Continental Shelf—the lowering of the sea level six hundred feet—would add millions of square miles of unexplored land and add profoundly to the geography lesson. This would make a land connection a thousand miles wide between North America and Asia! England and Ireland would have a dry union; and the possibility of sea sickness would be eliminated for travelers between England and France.

Perhaps the greater portion of the land area of all continents has been beneath the surface of the sea. Most of the sedimentary rocks once beneath the surface of the sea are segments of an old Continental Shelf. Through uplift this has been raised far into the sky. Although water-laid rocks carrying the shells of the sea have been found as the surface of mountain plateaus and summits, these were sedimentary rocks formed in Continental Shelves of shallow seas. No sedimentary rocks from the bottom of the deep ocean basins have ever thus been found. Apparently, the deep ocean basins through the ages have been but slightly changed.

The geological history of the earth indicates that its surface has gone through countless upheavals and subsidences. In places, these uplifts have embraced many thousand square miles. These have been pushed up thousands of feet, while other areas have been only slightly raised. There are many instances of small areas being slightly upraised or of their slightly sinking. As a matter of fact, it appears that, in a number of places, the surface of the globe has ever been both sinking and uplifting at the same time; and also, at the present time, a number of areas are either sinking or being upraised.

In most instances of even extensive upheavals or subsidences away from the sea shore, the record is not obvious and is not likely ever to be recorded. There are marked exceptions—for instance, the uplift of a mountain range.

A few million years ago, there was a mountain

revolution in the western part of what now is the United States. Like all revolutions, this overturned many prominent figures and lifted others into prominence. A stretch of Pacific sea bottom was uplifted and given a place in the sun.

An arid climate was the lot of most of this new scenery. This dryness was due to the uplifted Sierras intercepting most of the clouds on which the Pacific had been sending water supplies inland.

Out of this wild change the Great Basin came into existence, and most of the ages since it has been a dry basin except in spots.

In imagination, I can see the Great Basin come into existence and restore two vast and ancient lakes in desert scenes whose saline surfaces rose and fell amid the arid mountains with waves wearing and washing many a sandy, alkaline shore.

Mountains to north and south rise up and send the Snake and Colorado rivers on their ways to the sea; the Wasatch and the Sierra build barriers to east and west, and the Great Basin with detached ranges lies within the mountain rim.

Off the Sierras, the Wasatch and other mountain ranges, rivers poured down into the basin, but none of them poured out of it. For a million years and a day the Great Basin, with an area about equal to that of France, did not send a single drop of water to the sea.

The in-pouring streams do support a number of lakes, most of which are saline. Among these are Pyramid Lake, Nevada, and Great Salt Lake, Utah.

The waters of the Basin are shut off from the sea; the moisture-carrying clouds are nearly shut out from the sky. This land of little rain and much evaporation goes through dull ages. The impounded waters make little advance. The Ice Age comes.

There is a cooling, a check on evaporation, and increase of precipitation. Saline waters deepen and expand and freshen. A small lake overflows a pass and unites with another, then another. The water deepens, passes and islands become straits, the mountains seem to sink, and

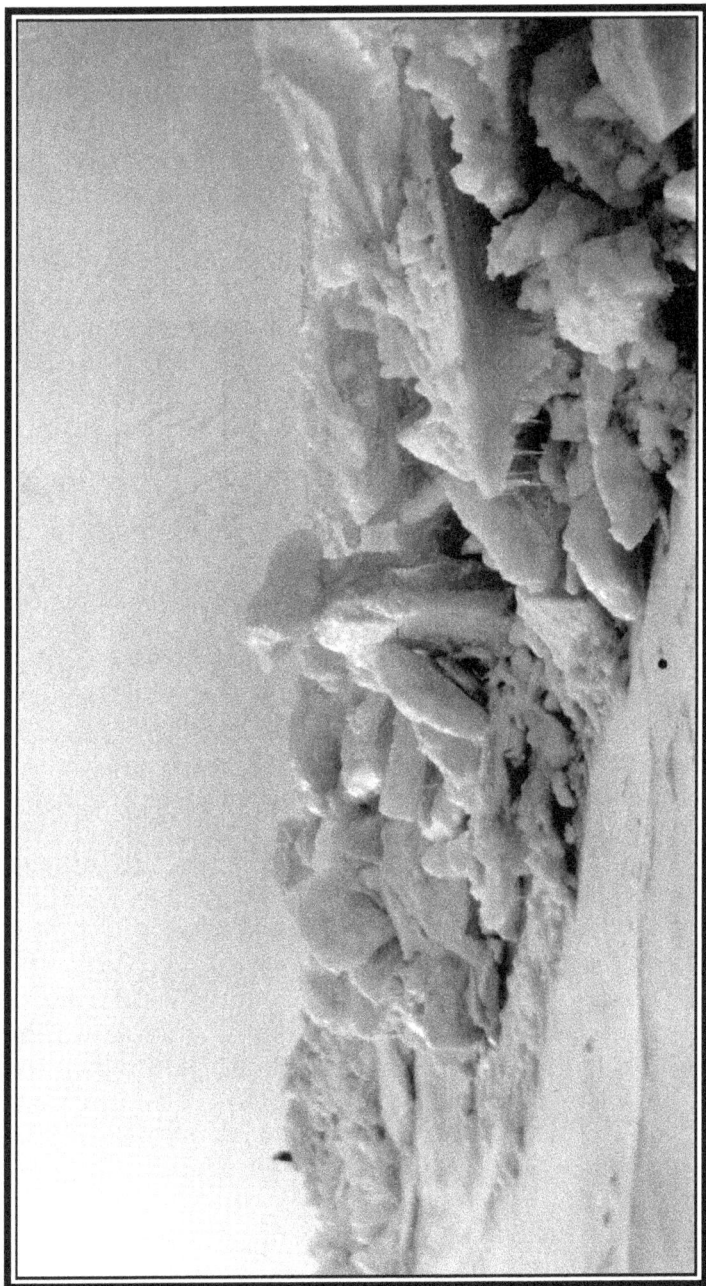

Ice builds up against the shore of a lake.

peaks stand out in this inland sea.

For a time, the waters cease to rise; they hesitate; they start to recede then rise again higher than before. Lake Bonneville, after piling up water to the depth of a thousand feet, is at last able to peep over a pass; it rushes over—escapes to the sea. Its outflow escapes through the Snake River and Columbia River. It cuts its outlet wider and deeper.

Its shoreline remains an age at the same level. Deltas build out into it. Beaches are worn, and shorelines cut into enduring rock. Its waters fall below the outlet; they try to reach it again and again. They give it up and slowly settle a few hundred feet; then pause and sink below the one-hundred-foot depth, then into its old sea bed. The tiny remnant of it is Great Salt Lake.

The bold, broad shoreline, beaten by the waves of Lake Bonneville thousands of years ago, show strangely, romantically, along the mountainsides.

Almost every movement in the Continental Shelf becomes significant. It is close to that profoundly important line called sea level. A few feet of uplift of the sea bottom, the Continental Shelf, will raise it above sea level. This will enable plants to grow over it and animals to live upon it. Wind and waters erode and sculpture it, leaving records of its place in the sun. Then, again, a slight subsidence of this area, and the bridge or thoroughfare which the area has formed suddenly ceases. Remains of plants and animals and its eroded surface may be covered with mud and shale and its fossil records thereby preserved.

The Continental Shelf has received, recorded and preserved a vast amount of geological lore. Nothing in the realm of geology offers so significant a story as do the various records of the Continental Shelf. During the long ages of the past, most of the sedimentary rocks have been developed from the delta material which has spread out on the Continental Shelf. In this shelf were found most of the fossils.

If it sinks slowly, peaty or boggy plants and trees which live in saline waters may grow upon it. This vegetation may slowly build up as the land slowly subsides, and thus keep

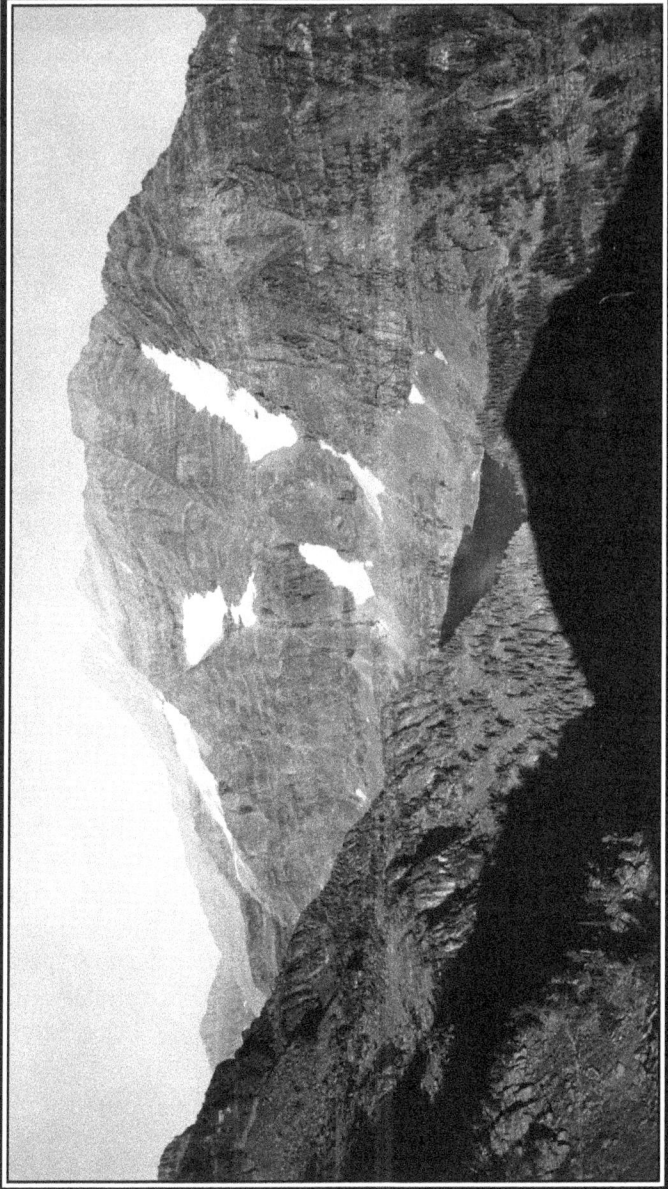

Mount Jackson from Lincoln Peak, showing layered striations built up over millions of years. Glacier National Park. Photographer unknown.

a growing surface at water level. It is thought that all the peat bogs and the coal deposits of the world have been formed in areas that were slowly sinking beneath the sea, by vegetation growing, falling, and piling up just as rapidly as the land slowly sank into the sea. In places, these coal layers are thousands of feet thick, showing a slow sinking of extensive areas over an enormous period of time. The coal marshes in Nova Scotia and Wales show more than sixty strata of coal separated by as many strata of shale rock. These would indicate repeated sinkings of land; then, the subsidence ceasing, shale building up to water level, or uplifts which brought the shale above water level, another growth of vegetation, and another sinking process.

Thus, in addition to the countless fossil remains swept on the Continental Shelf, the upheavals and subsidences of the Continental Shelf have afforded coal-producing plants the opportunity to grow, and the processes for their pressing and preservation. At the oscillating seashore, which has ever appealed to the imagination on surface; at the sharp boundary line where the continents and oceans meet and battle; at sea level, slightly above and slightly beneath—there have been enacted, through the ages, strange phenomena of which we have astounding records.

Apparently, for long ages, the continents of the earth were completely detached, each separated from the other by water barriers, so that the animal life of each developed exclusive of the influences of the other. This was the condition at the beginning of the Cenozoic era. A geological map of that time shows North and South America without land connections; and Europe, Asia, and Africa separated by water. Some time later, during the Miocene epoch, North America and Asia and Europe and Africa became united by broad land bridges. Even South America and Australia were but slightly detached. A little later, South America was united with North America. This uniting of land areas was followed by migrations and inter-migrations of animal life between the continents. The evidence of this is shown in the fossil records. Shortly after these land connections, the animal life that had been developing exclusively in one continent made its appearance in another.

The Miocene epoch is of late Cenozoic era and geologically is recent. Before this memorable epoch, most of the earth's surface appears to have been uplifted and eroded away several times.

The topography of the earth, at this time, was marked by mountains and plains not now in existence, and we have plains and mountains new. The river system was mostly different from that of today. Norther Europe, Asia, and America were united with a still more northerly continent called Holarctica. This now extinct continent appears to have had plants and animals of its own. It now largely lies beneath the sea, remainder fragments showing in Spitzbergen, Greenland, and perhaps in north Europe, Asia, and America.

Australia appears to have been without connection with the other continents for a few million years. As a result, she still has many of the primitive forms of mammals very much as they were when her last land connection with other continents was severed. Thus, many of the plant and animal forms in Australia today probably are similar to those that were on other continents a few million years ago.

The fossils of both land animals and sea life record that the continents were repeatedly connected, but in late geological times, these land connections between some of the continents were again broken.

The earth's crust had many ups and downs during the Miocene epoch. Parts of Germany, Belgium, New Zealand, and Australia were beneath seas not yet seven. Other shores and lands were sinking. Windblown desert sand was burying many regions, preparing fossils.

In America, volcanoes, earth faultings, upheavals and subsidence were preparing the scenes which we enjoy today. A segment of sea bottom called Florida was breaking through tropic waves, catching its first glimpses of the sky.

Plateau Province was rising where storm and sun, gravity and running water could shape its now famed scenes. The Colorado River was beginning to shape that splendid and stupendous colored earth sculpture—the Grand Canyon.

The Sierras were rising higher; the coast range was coming into existence, the Andes were going up, and the Himalayas attaining greater stature.

In Europe, the heaving earth was pushing up the Pyrenees and the Caucasus. The Alps then slept in formless stone beneath the level plain of Switzerland. Late in this epoch the plain began to swell and break; the Alps were born and rose high on the horizon.

For millions of years the deep sea basins have not been emptied; their position and their volume have remained about the same. The main continental land areas, too, have remained substantially the same for ages and ages. There have been countless minor changes; continents have been severed and united again and again. There have been upheavals and subsidences; the Continental Shelves have risen and sunk again and again, but the ocean basins and the continents have not traded places or very greatly changed their boundary lines.

The earth's relative proportions of land and water are now 71 percent of water and 29 percent of land. The average height of land is less than half a mile, where the average depth of sea is more than two miles. To make land and water areas equal, it would be necessary to lower the sea level about ten thousand feet and add this gained territory to the present land area.

During the height of the last Ice Age, there was a heavy ice cap over a vast area in the Northern Hemisphere. This covered northern North America, Europe, and Asia to a depth of a few thousand and possibly seven thousand feet. As the water stored in this ice had been withdrawn from the sea, this stupendous quantity must have measurably lowered all the shorelines of the sea. When the changed condition allowed the warm sun to release this water stored in ice, its return to the sea must have raised the sea level, the shorelines, of the seven seas.

It is estimated that sediments forty miles deep have been deposited into the sea, and as layers of these sediments are of great thickness in places, this would indicate that a sinking of the shoreline depth at the mouth of a river is not

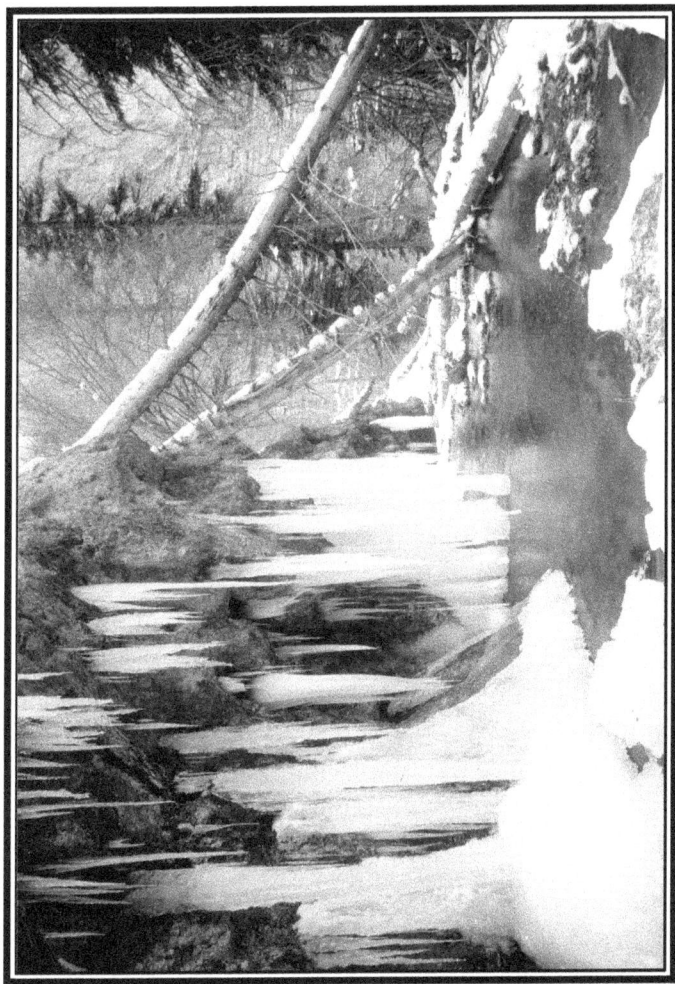

One of the walls of Hanging Lake in winter.
Photo by George L. Beam.

large.

Somewhere the sea is breaking through a section of its shoreline. Three times in less than a century the sea has compelled the building of the Cape Charles lighthouse several hundred feet farther inland. The former port of Ravenshire, England, no longer exists, and the sea surges where it long stood. In seventy years, Sharp's Island, in Chesapeake Bay, had been reduced by the sea to fifty-four acres. Helgoland, North Sea, has a diameter of three miles. The first accounts of this island give it an area about ten times its present one.

River erosion of land may be called vertical, deep-digging erosion, while erosion of the seashore may be called horizontal erosion which works inland. A river may be called a narrow, flexible saw which, with a cutting edge of flowing sand, cuts vertically into the earth. The sea, where it comes in contact with the land, may be called a long saw which cuts horizontally into the land.

The unequal hardness of the rocks gives an unequal, uneven shoreline. A land surface shows an uneven wearing away of the rocks because of unequal hardness and because of unequal distribution of water over them.

The shoreline of the sea may be compared to the skyline of the land. Place a rocky mountain range on its side and you have the bays and peninsulas—the indented shore—of many seas.

In southern South America numerous sea shells have been found on mountains more than one thousand feet above the sea level. In the Appalachian Mountains, rocks which have been beneath the sea are now high up in the mountainsides. Sedimentary rocks of sea formation are found in the Rocky Mountains ten thousand feet above the sea level, in the Andes three miles above, and in the Himalayas at still higher levels.

In a few cases the deposits have not been disturbed, and in other places, the strata have been formed, uplifted, partly eroded away, other strata formed upon the realm, and these again uplifted and eroded. Thus, the imagination, in looking out upon the earth through enormous periods of time, may see change succeed change, landscape melt away and

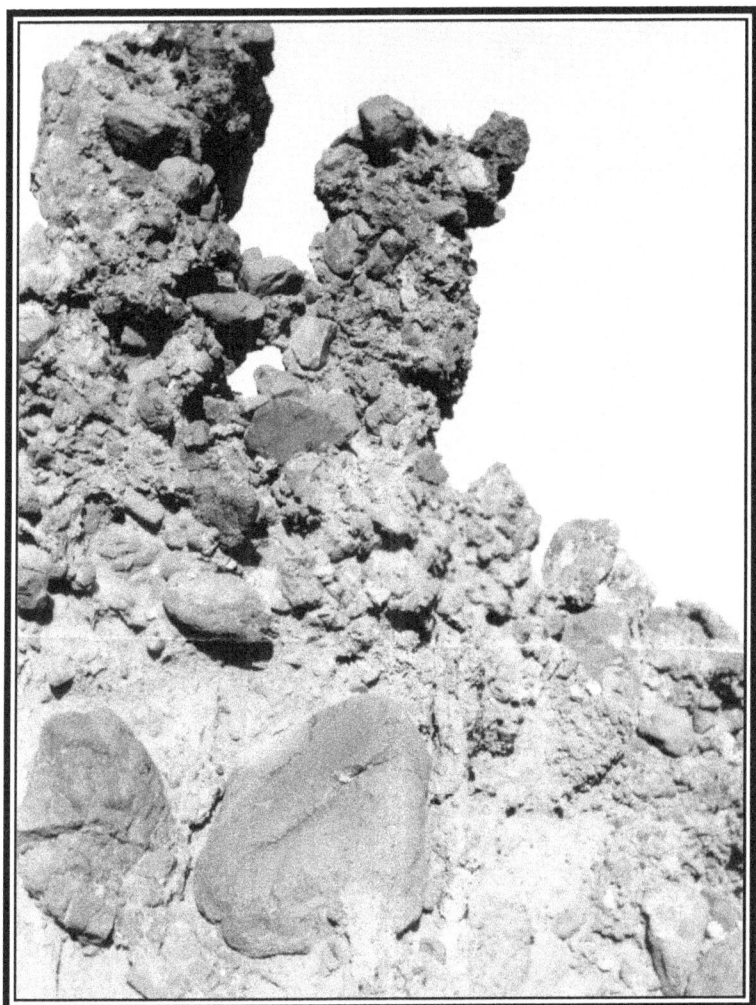

Eroded forms on Specimen Mountain.

a new one appear from the ceaseless action of that artistic sculpturing element which we call erosion.

It is the fourth appearance above the sea of the plateau region through which the Grand Canyon is cut. This time it has been in the sun long enough for several thousand feet of the sediments, which were deposited upon it during its latest submergence in the sea, to be eroded away. All these are uplifted continental plains or shelves.

Sections of river channels in places show near the summits of the Sierras. At one time, a narrow sea extended northward from the Gulf of Mexico to the Arctic, cutting the Continent in two.

Often, a new landscape is uplifted into the sunlight. No scene is ever completed. So slowly the waters rush off and the views conceal and delay that we cannot realize that the scene as we look is fading even faster than color from a sunset. Sometimes landslide, cyclone, earthquake, flood, fire or volcano shifts a scene before our eyes. Sometimes a skyline vanishes. The horizon that had its place against the evening sun melts away as the earth circles in shadow toward the east, and in its vacated sector a new landscape shows in silhouette against the morning sun.

Transient sand dunes, crumbling ruins, and ever-shifting scenes are a part of the thought-compelling landscapes of the earth. Their mysterious grains of insoluble sand are from all ages, from all the continents and the seven seas, and with them the imagination sits down like a child to play, build, and think.

Touched by the wind, the unresting sand dunes give a restful margin to the shorelines of all the restless seas. With the wind, they gipsy upon the bends and banks of every river, play in wild abandon over leagues of desert, and around ten thousand dreamy lakes and ocean beaches maintain the wondrous, immortal No Man's Land, the land where children play and dream.

The play ever goes on. Sediment in Continental Shelves may not long play its part at the bottom of the sea; up-piling will give pressure that forces the bottom to flow out and up. This stratum will in due time move up and on to play again its part in light as peak against the shifting cloud scenery of the sky, a towering sundial point with league-long shadow moving opposite the sweep and swing of the sun.

Geodes on Specimen Mountain.

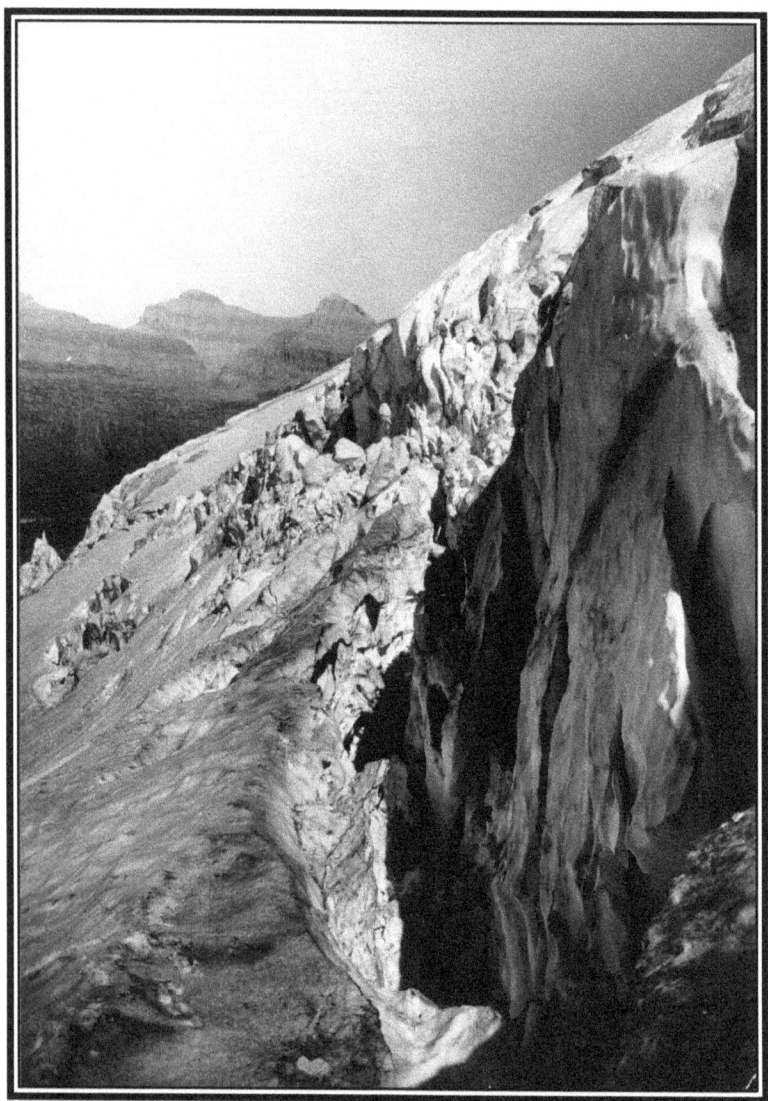

Blackfoot Glacier, Glacier National Park, Montana.
Photograph by Fred Kiser.

Running Down Glaciers

For many years, the mountain people of Switzerland asked the question, "Do glaciers move?" Then a glacier without warning came down out of its canyon across a meadow and pushed a barn over. That glacier had its proper standing. A little later, another glacier moved down and chased the workmen out of a marble quarry. Elsewhere in the Alps, about the same time, a man built a small hut upon a glacier. In due time this hut was moved forward by the ice and traveled with the glacier more than a mile. These incidents, together with the measurements of scientists, settled the question. Glaciers move.

The entire southern coast of Alaska has been shaped by glaciers, and when I visited Alaska in the summer of 1892, any number were still sliding from the steep, snowy mountains into the sea. These young bergs, many of them of enormous size, rolled ponderously about, bowing profoundly and making a stir as they entered the world-round sea.

I was looking for glaciers, and here seemed to be a splendid world's fair exhibit on them. While watching one from close range, it shed off a huge berg. This bobbed up and rolled heavily over. The high wave that it sent far up the shore threw driftwood about like a flood, and carrying me along with it, finally pitched me headlong at the line of high tide.

Coming upon a big berg becalmed in a small harbor with its snout inclined against the shore, I concluded to climb up. It rose fifty or sixty feet above water, and had a plateau surface. A few steps had to be chopped, but with little difficulty I reached the top. It was nearly level and about four hundred feet long, though less than one hundred wide. In one spot were several well-rounded boulders like big eggs in a nest. These were ready to roll overboard. A flat triangular-shaped stone was supported several feet above the common level by a triangular ice mass which the stone had shaded and thus delayed its melting. A number of

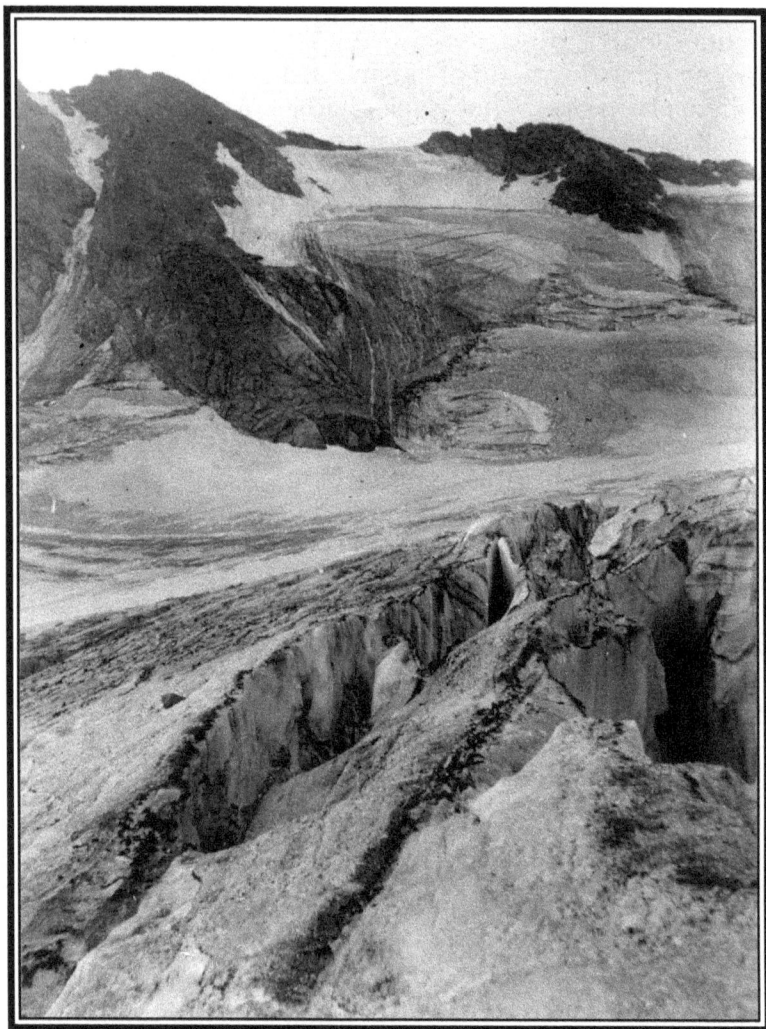

Crevasses in Arapaho glacier, September 1, 1902.
Photo by Junius Henderson.

boulders like large peanuts were set in the ice. A few carloads of sand and gravel were frozen on and in the offshore end. Apparently a gravel-laden stream, while the berg still had a place in a glacier, had poured down upon it. There were no deep cracks, and if there had been ancient crevasses, these had been closed and mended to as to defy detection.

What most interested me was a pile of confused logs partly embedded in the top and one side of the berg. These were mostly of broken spruce trees and probably had been swept down upon the glacier by a snow slide. I had carried my bear skin and another blanket up with me, so I concluded to have a night at the top. The logs burned well, and my rousing fire lighted the spruce-walled shore. Knowing that a heavy swell or something else might start the berg on an adventure out to sea, I was careful to watch for any unusual movement, and ever listened for sounds of rising or surging waters. Perhaps the chief reason for spending a night on this iceberg was Youth.

Just how many billion snowflakes were compressed in this ice cake could not be figured. If it had been assembled and started from the extreme wing of the canyon down which it had apparently slid, and had slid forward 1,000 feet per year, the journey to tide water had required eighty-four years. Its descent from source to sea was about nine hundred feet.

Many a glacier never reaches the sea, and may be likened to a desert stream that evaporates and disappears in the sand. In a wet year, such a stream flows beyond the ordinary place of disappearance, while in a dry year it does not reach this place.

With a big fire burning, I lay down for a sleep. I had figured that I could get off the berg quickly in case it suddenly started out to sea. Of course, I might have to swim, but the shore was close. Strangely, with marvelous distinctness shone the stars, and I was alone, a stranger in a Past of several thousand years ago.

Though chilled, I awoke with the fires of sunrise blazing into the sky behind the black mountain wall. The berg's restlessness had awakened me. When I was less than a quarter of a mile along the shore, the berg did put out to sea.

There are only a few small glaciers in the Rockies, but small as they are, a visit to their blue-green ice piles in midsummer stirs the imagination. The number of glacier meadows and the numerous well-preserved old moraines easily enable one to form a picture of the region during the Ice Age. Most of the canyons in the Continental Divide which lie in altitudes of from eight to thirteen thousand feet have glacier-polished walls.

At one place in the Rocky Mountains, I saw a small glacier at an altitude of 13,000 feet, which was evidently in a wind eddy near the summit of the range. Apparently, it was maintained entirely by the snow which the wind blew to it. This snow had come from off an area above the timberline which aggregated about four square miles. In this and other instances, a change of direction in the prevailing wind, and the glacier would vanish.

In the Rocky Mountain National Park, the few remaining glaciers are on the east side of the Continental Divide. On the west side of the Divide there is a heavier snowfall, but there are no glaciers. The cause of this condition is prevailing westerly winds. These sweep much of the snow from the upper western slopes and the summit and deposit it in the upper ends of the canyons with the "eternal snowfields" on the eastern side.

Commonly, a glacier is formed in the upper end of a canyon by the vast quantity of snow which slides down off tributary slopes and is swept to it by the wind—even the snow which blows off the other side of the mountain.

Wherever more snow falls or accumulates each year than melts, an icy mass will result, and in due time there will be a movement of this ice. The steepness of the slope will determine the quantity which is required to produce a movement. On a steep slope, a moderate quantity will move. Weight and rapid melting compact snowdrifts into ice. Ice under pressure becomes plastic, and where a few hundred feet are up-piled, the weight of the top is too great for the bottom layers to withstand, and it is squeezed out.

The main motive power of a glacier is the down-grade

Above: Neve remnant in cirque north of Arapaho.
Below: Arapaho Glacier from the south, July 29, 1904.
Photos by Junius Hunderson.

pull of gravity. The expansion and push due to freezing also advance it. The pressure of the up-piled weight at the source is also felt all along the line.

Glaciers—ice piles—move for the same reason that causes snow piles to slip and slide, because they were dropped or formed on a slope so steep that they were ever slipping.

On the upper end of a glacier in Glacier Bay, I found a mountainous mass of snow, near-ice, piled upon the glacier. This accumulation of several snow slides, one or more of which appeared to have come down each of the several gulches, all united on the glacier. This was the source and resource of the glacier.

This mighty pile rested upon a foundation of glacier ice which filled the canyon below. This foundation could not support its weight of snow and ice, and constantly slipped and slid forward down the canyon. Storms, winds, and snow slides maintain the glacier reservoir, supply the pressure, and thus sustain the flow—the downward, forward movement of the glacier.

In a number of places, glaciers have traveled part of their way up grade, up a slope and over a ridge. The inconceivable pressure of the enormous and elevated masses of ice at their upper reaches and sources had forced the flow forward.

In the Yosemite National Park, the ancient Tenaya Glacier flowed westerly from the summits of the Sierras. It left a strange and impressive story of the ways and the work of ice. It was about two miles wide, more than fourteen miles long, and in places two thousand feet deep. Between Tuolumne and Tenaya basins, it slid up a steep slope, climbing five hundred feet, received a tributary, then poured over against Clouds Rest and down upon the domes of the Yosemite.

What a wild, grand sight this rough, wide, deep ice stream must have made as it swept onward and downward from the summit of the Sierras! Mount Hoffman, miles in front of the range, split this grand ice stream as a rocky island splits a river. One fork of this ice veered off to descend and work out the sculpturing in Hetch-Hetchy. The

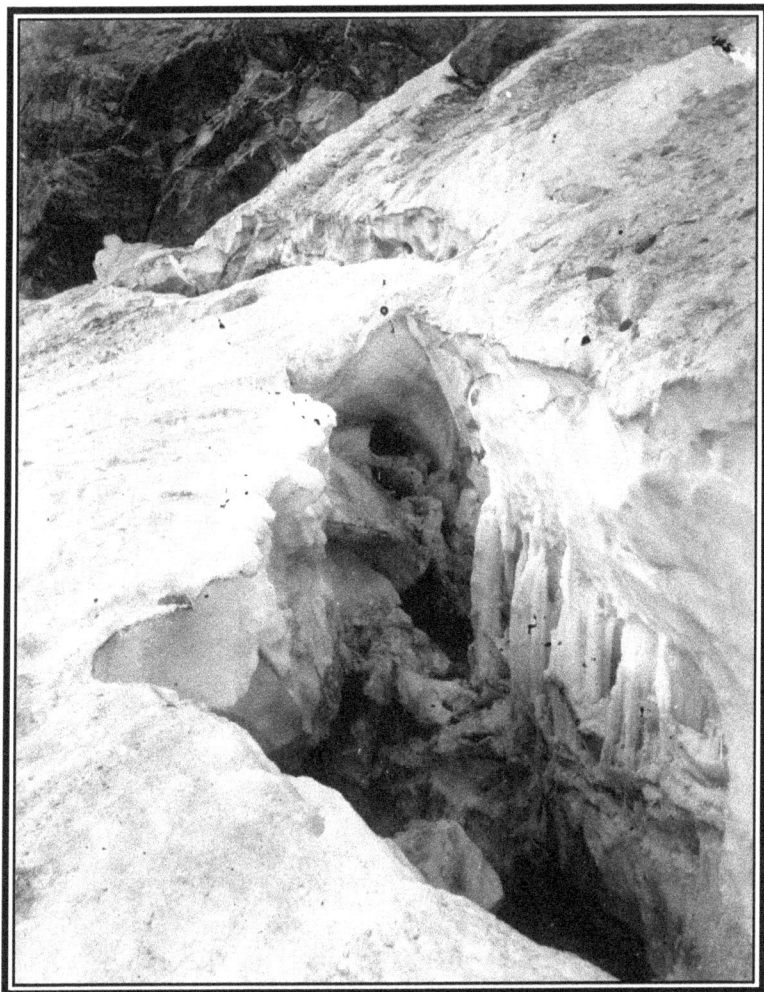

The Bergschrund, Arapaho Glacier, September 1, 1902.
Photo by Junius Henderson.

other gave its attention to the Yosemite, climbed up a steep, was joined by a tributary stream, where it went over the top, and then descended to help shape the stupendous and splendid rock sculpture through the length of the Yosemite Valley.

The weight of a glacier, especially along its upper, deeper reaches, is enormous. Usually, the upper part is laden with large quantities of rock and landslide material. The vast up-piled ice and stones cause the glacier to bear heavily on the bottom of the canyon and to exert against its walls a telling pressure, sometimes aggregating one hundred tons per square foot.

This rough mass of ice and stones, moving under enormous pressure, enables it to cut all touched surfaces away. Its erosive force is enormous. The mass of ice mingled with rocks moves slowly forward, tearing off projecting rocks from the walls and from the bottom of the canyon; it rasps, planes, and polishes all surfaces touched. When a glacier continues to flow for thousands of years, as glaciers have frequently done, it greatly deepens and widens its channel, smooths and wears off the surface over which it travels. Rough V-shaped canyons are smoothed, deepened, and made U-shaped. Mountain peaks are planed down to low turtle outlines, and low-lying hills are leveled off into plains.

The glacier's erosive power appears greatest headward. A glacier sometimes temporarily freezes fast at the source; a lurch and forward movement, and the frozen upper end drags out rocks of many tons. It thus works its way headward into and often through a mountain. Had the Ice Age lasted another century, Long's Peak might have lost the remaining part of its head.

The nose or front of a glacier made up of ice, and ofttimes an equal quantity of stones, has a formidable appearance. It is effective in ploughing and cutting its way. Where a glacier descends a steep slope and comes in contact with a level stretch, it commonly bears down and cuts deeply, thus forming a basin of solid rock, which, after the melting away of the ice, is filled with water. Most mountain lakes occupy basins that were formed in this manner. The Rocky Mountains, the Cascades, and the Sierras owe their numerous lakes to glacial action. The most beautiful lakes in New York State, in the Glacier National Park, those in Scotland, and, in fact, the over-whelming majority of the lakes of the world, repose in rock basins which were scooped out by the slow, vigorous action of glacial ice, or they rest in reservoirs that were formed by glacial moraines damming a section of a former river channel or a previously cut canyon. Hudson Bay was probably excavated by glaciers.

Wisconsin is dotted with lakes of glacial origin, and many of its lakes repose in depressions of glacial drift. The Great Lakes are glacial. A number of the long, narrow lakes of New York State are parts of old river channels. Of course, a few of the lakes of the world were formed in basins produced by landslide debris, clogged river channels, or in the former fiery and abandoned craters of volcanoes.

In the Rocky Mountains there are innumerable meadows and valleys in which flourish forest, grasses, and flowers in the soil ground for plant food by the glaciers. This soil comes chiefly from the ground-up rock—the rock flour—which results from the grind of glacial movement. Possibly one half the soil now serving over the earth is traceable to glacial origin. Soil was in part distributed by glaciers and further outspread by the action of wind and water. Many of the more extensive and productive grain-growing areas in the United States have glacial soil. Many farms districts in Canada and Europe are in glacial drift. And thus we may say that one of the prime resources of the earth—soil—which now makes the earth livable, is the great rock grist which the glaciers ground and transported and which wind and water and the chemistry of nature made ready and widely distributed.

But in addition to these activities, glaciers have given flowing lines to landscapes, have beautified the earth with rounded hills, and have decorated it with lakes of exquisite beauty and water basins of every form and size. A majority of the Sierra landscapes are new, recently made by glaciers, and nearly all the forests in the Sierras and the Rocky Mountains are growing on glacial moraines.

Above: "Arapaho Glacier at its best, Sept. 1st, 1902.
When, after a very dry season, snow was melted
back clear to the neve, exposing all the crevasses
and the Bergschrund."
Below: Fair Glacier, September 1, 1910.
Photographs by Junius Henderson.

John Muir long ago pointed out that the gentle, delicate snowflake—a snow flower—is a rock trimming and polishing agency of importance, an earth artist of first magnitude. When one thinks of glaciers he must necessarily think of landscapes, soil, and scenery.

In a moving glacier, its front and bottom and sides are set with innumerable cutting tools in the form of broken rocks. These rocks wear out or become dulled, and as they advance in the ice of the glacier, are dropped in the lateral moraines or dumped at the terminals. The ice has continuous flow and the landslides from the heights are frequently dropping whole trainloads of new tools—broken rock—upon the glacier. Then, too, the glacier is constantly seizing rocky material at the head or tearing it from the sides and bottom of the canyon. Running water, too, covers and fills ice full of sand and gravel. This also helps to cut and polish the floor and walls of the canyon.

After a glacier has flowed through the same channel for a thousand or more years, it has measurably straightened, widened, and deepened this channel. The ice may then melt away, but its enormous carving remains, and in the lateral and terminal moraines and in the outspread silt drift and soil may be had a glimpse of the material which it moved.

On one glacier, vegetation was growing in the soil of a rock garden. This showed a number of flowers in bloom, alpine gentians, yellow avens, and purple primroses. On the edge of a crevasse nearby rested a number of arctic ptarmigan. A cony was squeaking among the rocks, while on the nearby ice a number of rosy finches were feeding.

In places, I have found wild animals well preserved though long dead in the terminus of a glacier. They evidently had fallen into crevasses possibly a hundred or more years before, and has lain in cold storage through all the years. Most material that falls into crevasses is likely to be crushed or to be ground up as the ice advances over its uneven bed.

A glacier, like a river, transports vast quantities of material, mostly rocky debris. This usually starts in the form of rocks, which are ground to flour, pebbles, cobblestones, and boulders. The material eroded and carried forward by

"Glacier National Park, Montana, shows strikingly
the work of the Ice Age, in its ice-carved peaks, glacier-
made lakes, and large remnants of the age-old
glacier still remaining."
Looking northwest from Castle Mountain.
Photo by Fred Kiser.

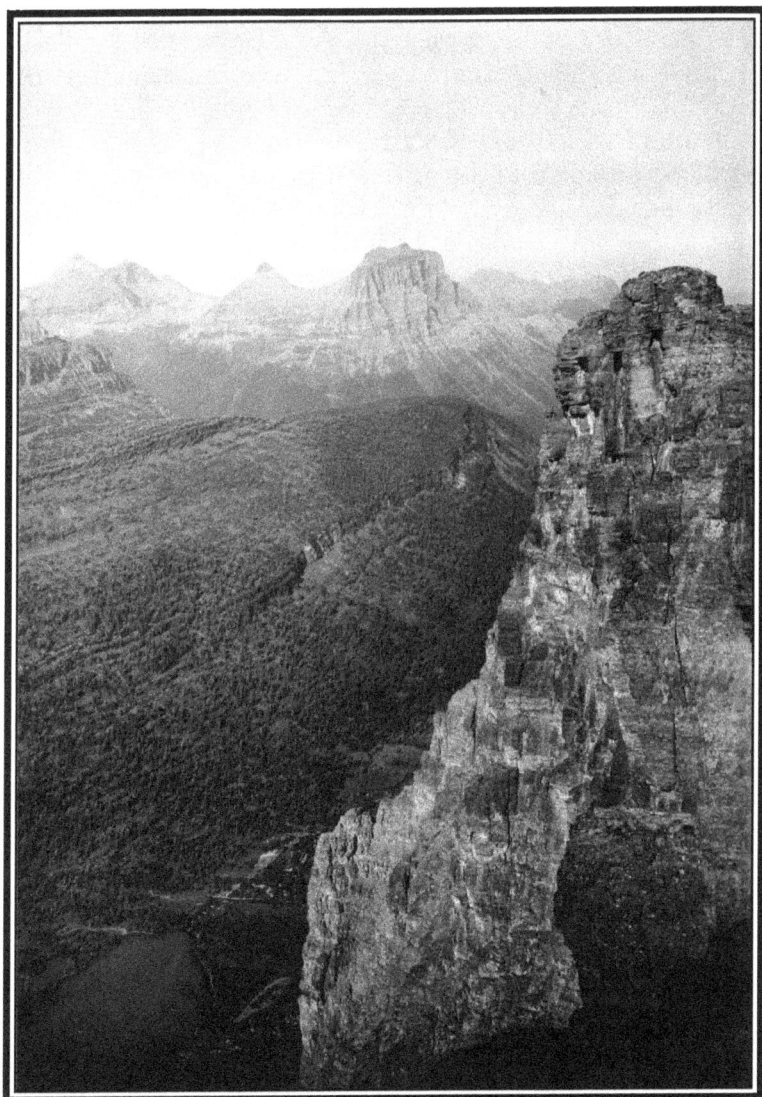

Looking northeast from the summit of Fusillade
Mountain, Glacier National Park.
Photo by Fred Kiser.

the glacier may fill in canyons over which it is passing or finally be dumped at the terminus of the glacier. At the lower end or terminus of a glacier, where all the ice melts away, quantities of this rock debris is dropped, and is called a terminal moraine. In many a glacier, this would amount to hundreds of tons daily.

Terminal moraines of vast size may be seen in the mountains and also in the Mississippi Valley. Much of the area of Iowa, Dakota, Minnesota, Wisconsin, Illinois, Indiana, and Ohio is overlaid with glacial drift and morainal matter from ten feet to five hundred feet deep.

The topography of the greater part of the Northern Hemisphere has been carved chiefly by glaciers. Canada has undergone vast and inconceivable changes from ice erosion, and the northern surface of the United States was wrought mainly by glacial forces. They have cut thousands of mountain canyons, enlarged and reshaped others, dressed mountain ranges down to plains.

By dumping or depositing vast quantities of morainal debris, glaciers sometimes fill a valley and extinguish a stream or change the direction of the water flow.

The Missouri River formerly flowed northeast and perhaps emptied somewhere to the northwest of Lake Superior. This lake went out of existence, and the Missouri River was broadly pushed a few hundred miles to the south, and its waters made connection with the Mississippi River.

Long Island is a terminal moraine, the delta of an ice river that was piled during the last Ice Age. The material was scraped and transported from the mountains of New York, New England, and Canada. This stupendous mass of material will give some idea of the inconceivable quantity of debris which glaciers accumulate, transport, and distribute.

A glacier may terminate—melt away—at about the same place year after year and pile up an enormous terminal moraine. A few years of scanty snows, and it will retreat; that is, its lower end will melt away without reaching the old terminal place. If there comes more than ordinary snowfall, the glacier will in due time respond, and its end plough through the deposited moraine delta and push beyond the ordinary terminal point.

If a glacier moves five or six feet per day—about two thousand feet per year—terminating each year at the same place, two thousand feet of it will melt away each year, and the weight or quantity of rock debris deposited each year will be the amount in the terminal two thousand feet.

After the glacier emerges from the canyon, the rock material in the sides of a glacier and on its top often rolls off and forms great embankments or levee-like ridges— lateral moraines, on each side of the ice stream. The Rocky Mountains and the Sierras carry thousands of these bouldery moraines, which in places look like extensions of canyon walls.

A boulder may have had strange, violent experiences. The original rock fragment may have been torn from a cliff which projected into the canyon, from the bottom or wall of the canyon, or have been plucked from the uppermost end of the canyon. It may have fallen from a skyline cliff and tumbled down upon the glacier where, for a time, it was carried on the surface. Later, it may have dropped into a crevasse and reached the bottom. Here, wedged in the ice, its sharp corners and edges may have been brought in contact with the rock over which the glacier was sliding, and it may have gouged and rubbed against this under a pressure of one hundred or more tons per square foot. It was crushed, ground, rolled, scoured, and carved. By coming in contact with other rocks receiving similar treatment, it may have been pushed to the surface at the top or the side, to roll out upon the lateral moraine, or to become one of many in the terminal moraine.

The more spectacular America and European glaciers are long, narrow tongues of ice, commonly in a canyon. They may be one hundred feet or a mile wide, a mile or several miles long, and from a few feet to many hundred feet thick. Many of the existing glaciers are long, narrow ice rivers occupying gorges in the mountains.

The greatest glaciers on the continent are in Alaska. Greenland still is largely covered with an enormous glacier, and in the Antarctic, glaciers cover thousands of square miles to vast depth. The largest glacier in Switzerland is

Above: Front of Arapaho Glacier showing stratification
and hummocks, September 2, 1903.
Below: Part of moraine of Isabel Glacier.
Photos by Junius Henderson.

Above: Isabel Glacier. Below: Arapaho Peak and its
glacier, south of Isabel Glacier.
Photos by Junius Henderson.

about ten miles long. There are glaciers in New Zealand, in the mountains of Asia, in the Andes, and one in Africa almost beneath the equator. Mount Rainier has splendid glaciers on its slope, the aggregate area of which is about fifty square miles. There are scores of glaciers in the mountains of Canada, numbers of small glaciers in the Sierras of California, in the Rocky Mountain National Park, and in Glacier National Park.

All over the earth, for many years, because of lessened snowfalls, most glaciers have been retreating; that is to say, melting back at the lower end a little faster than they move forward. Here and there are exceptional conditions, perhaps increased snowfalls at its source, which cause a glacier to lengthen or advance.

The rate of movement in glaciers varies from a few feet per month to several feet per day. Generally the larger the glacier and the steeper its inclined channel, the faster it moves. In rare instances one may move forward twenty-five or thirty feet in a day.

Some tourists were one day looking at the terminus of a glacier and the guide was explaining that the glacier was moving forward a few feet each day—and also that the terminus had been at the same place for several years. "Evidently, then," said one of the tourists, "it is always moving, but never gets anywhere." A glacier travels gracefully. It is ever grinding soil and making landscapes.

There appear to have been at least five great Ice Ages during the long history of the earth. These have been levelers and have produced inconceivable topographical changes. The Great Ice Age, the most recent of the five, markedly changed all the northern part of North America.

It was in 1840 that Agassiz brought forward the Ice Age theory—showed the recent glaciation of norther Europe and America. For nearly a generation, this now apparently obvious theory met with opposition. It is now almost universally believed by scientists.

One geologist has estimated that a permanent lowering of the temperature of five or possibly ten degrees in the temperate or arctic zones might bring on another Ice Age; and, too, a slightly increased snowfall in the north temperate zone might multiply the number of glaciers and increase the activity on the existing ones.

Glacial lakes below the terminus of Arapaho Glacier.
Photo by Junius Henderson.

Earth's Big By-Product: Soil

A sand bar in the Mississippi may be called the fat of the land. It contains an aggregation of adventurous soil, contributions from Iowa farms, fragments of Ohio hills, rock flour from the Rocky Mountains, golden sand from the Yellowstone, brown dust from Wyoming, and volcanic ashes, windblown from craters. These are combined with mineral salts, the sweepings and the scum of many rivers, and all form a rich soil that is eager to produce corn or grass, trees or flowers.

Though soil is composed chiefly of rock, it is a combination of rock dust, chemicals, and organic matter. It may be made from marble, coal, or volcanic lava. A little of almost everything may be in a yard of soil, and numerous are the forces which had a part in its compounding. Dust, quantities of it, falls upon the earth from other worlds; dust from shooting stars becomes soil!

With ten boys and girls I examined the ore and mineralized quartz piled by the top of three old prospect holes. The children pocketed a number of pieces of the shining peacock-colored quartz. Then we climbed down into one of these shallow holes, which was about four feet deep, that had been blasted into rock. All were astonished that the thin coat of soil surface, like a coat of dark-gray paint on the solid rock, was less than one inch thick. They felt it and pried their fingers between it and the bedrock upon which it lay. They were excited over this glimpse of geology, and wondered if everywhere solid rock was just beneath the surface of the soil; so we climbed two miles to where miners had blasted a tunnel into the mountainside. Here again was solid rock with just a thin green-gray blanket of soil covering.

"If we were to extend this shaft through the earth," I told them, "we should not strike any more soil until we came out on the other side; in China, perhaps."

As we stood by the entrance of this tunnel I said, "I wonder if you boys and girls realize that, once upon a time,

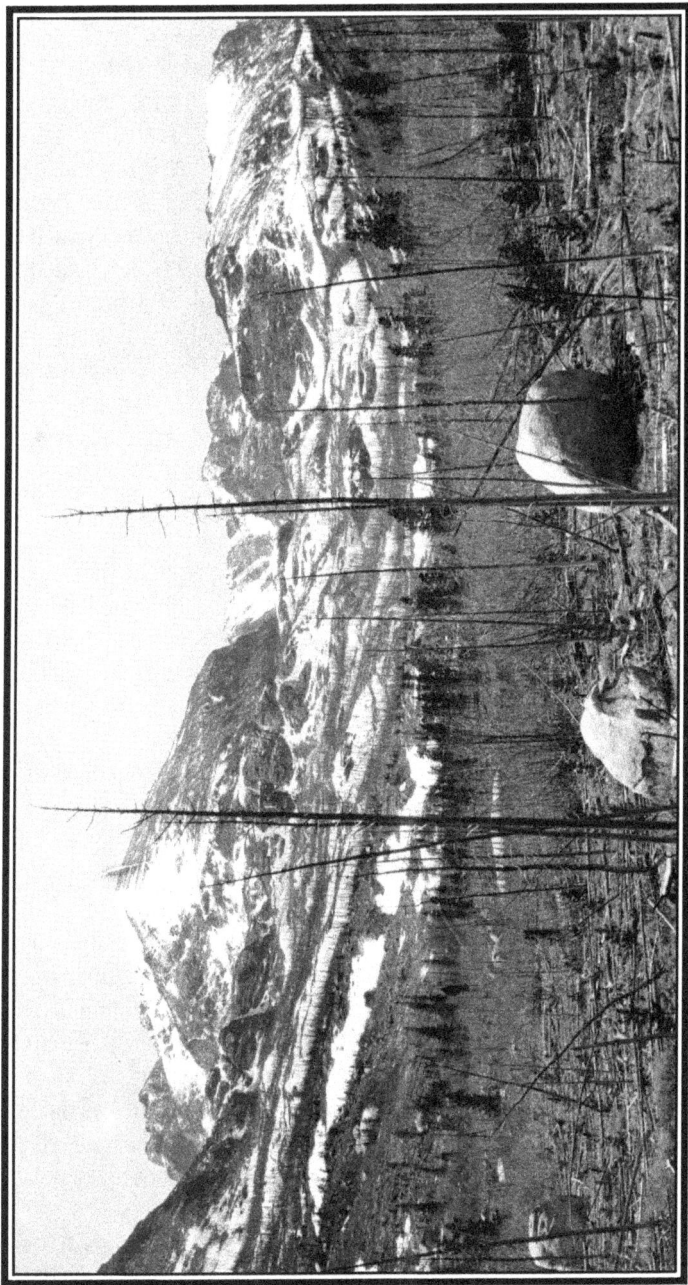

The greater part of the forests of the Rocky Mountains are grown in rich glacial drifts or deposits.

many million years ago, this round earth was solid stone; not a bit of dust or mud or soil upon it; no green grass, no pines, none of the colored flowers, no bears, fish, or butterflies. Stone has the quality which enables it to go through more changes than any other thing that we can think of. Through the action and interaction of chemical and other forces of nature, the solid rock surface of the earth has slowly decayed and formed a thin covering of soil. Over wide areas of many hilly and mountainous sections the naked rock is exposed. Soil has not yet decayed in sufficient quantities to cover these places; or the soil is blown or washed off the steep slopes almost as soon as it forms. A bare inch of soil covers other regions. But there are numerous valleys where the soil may be a few hundred feet deep. This has been carried into the lowlands by the waters, and may have come from off hundreds or even thousands of square miles of higher regions."

The children were from several states, and all eagerly agreed that, when they returned home, they would dig into the earth where something was growing and let me know how deep the soil was. All but two found rock less than a foot below the surface. The Illinois girl struck a boulder at fifteen inches, and the boy in New Orleans soon found water beneath the sand.

Later, one of the boys visited a recently drilled oil well and wrote: "The drill first went through several inches of soil, then struck rock, and went through layer after layer of different kinds of rock. Though down more than half a mile into the globe, it was still in rock."

Soil is a magical resource. Without it, the earth would be lifeless; with its life- and growth-giving power human life and all other kinds exist comfortably in a mysterious and beautiful world. The vast forests—the great and splendid trees that stand like fixed pillars while generations of men pass by, the grassy plains and prairies, innumerable fields of tasseled grain all golden in the sun, the orchards, the myriads of flowers with color bloom shining—all these grow from life-producing soil.

The ocean bottom has its productive soil. The strange and luxuriant plant life that waves and nods beneath the surface of the sea; the entangling vines and the weird forests that flourish where sharks and other great sea fishes swim; growth of green moss and moss of rosy tint beneath which little fishes and small folk hide; kelp and other plants queer enough to belong to the desert flora, the wondrous growth in coral caves—all these are sustained by soil, both solid and in solution, that originated from the rock surface of the earth.

It is said that nature requires ten thousand years to produce an inch of soil. Rock—the original raw material—is cut and dissolved by water, worn by wind, ground to flour by glaciers, wedged apart by ice, shattered by oft-changing heat and cold, and crushed to powder in falling.

An echoing on the cliffs opposite me was my first warning one day that a rock slide was coming down. It was an avalanche of several thousand tons plunging off the slope of Mount Lady Washington into Chasm Lake. It was accompanied by the whiz of falling stone. A huge stone struck and pierced twenty feet of snow and more than four feet of ice which covered the lake. Then a trainload of rocky debris plunged into the lake, throwing tons of ice and water wildly flying. The report of the striking repeated and reechoed by surrounding cliffs and steeps made an uproarious crashing, as though the top of Long's Peak had collapsed.

Wherever rock is exposed—in peaks, cliffs, boulders, and pebbles—its surface is alternately heated and cooled again and again. A thin layer, through expansion and contraction, scales or flakes off, and the expansion and contraction of a new surface takes place. Water, air, and chemicals act on these scales—this rock waste—and change them into soil. In many localities a number of soil-making factors are working together, but in places they work singly.

Air and water on the surface of rock combine with the mineral acid of the rock, and this sets up a chemical action that eats into and dissolves the rock surface. This reduction of solid rock is extremely slow. A cracking of the rock allows air and water to penetrate deeply, and this increases the exposed surface. Crushed and powdered rock is, by the presence of air, water, and acid, speedily made ready for

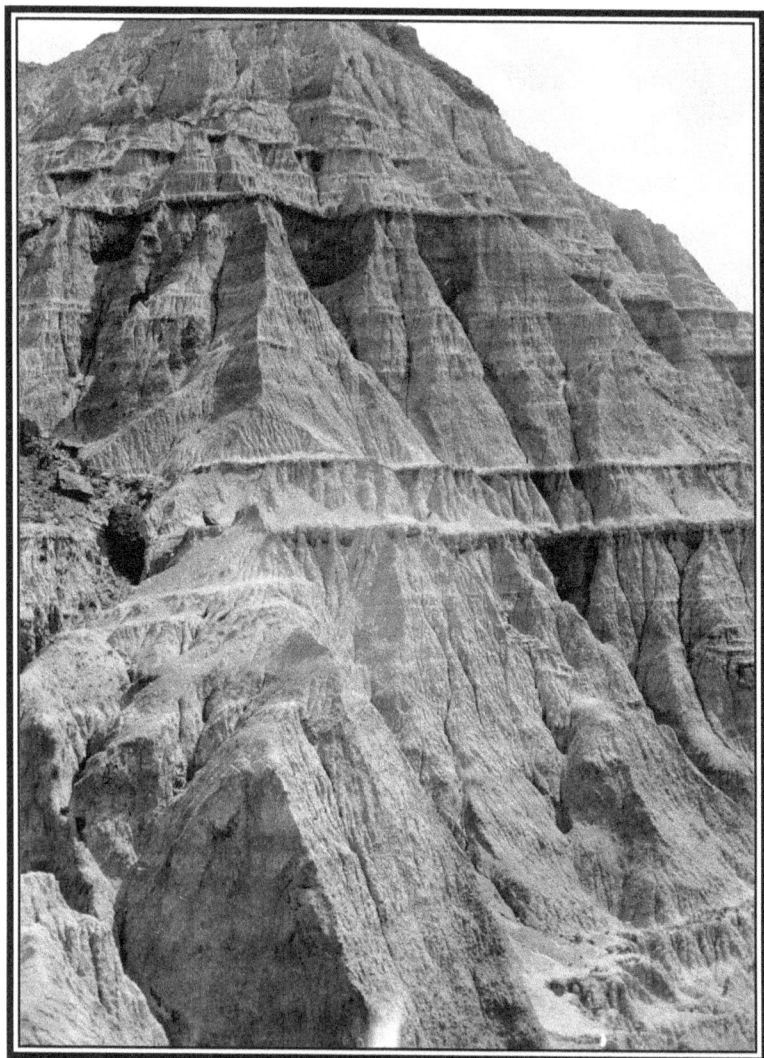

Rain and wind sculpted soft tertiary lake beds at the head
of Camp Creek, near Price, Oregon.
Photo by Russell.

soil. As time goes on, the general shattering, cracking, and roughening of rocks by the erosive forces render them more exposed and more readily susceptible to the chemical change that prepares them to be producing soil.

I saw a high cliff fall more than one hundred feet and land upon solid rock. How many years had water undermined and ice wedged to throw it over? It was shattered to fragments, and several truckloads of rock dust were at once spread out to begin life as soil. Young spruce threes were soon growing here. The fall of rocks in mountains is every going on, reducing ounces, pounds, and even tons of rock to powder-near soil.

One spring day, I saw a landslide in the San Juan Mountains. An enormous mass of earth and rock slipped out of the side of the mountain and slid down the steep slope, spreading itself over an area several hundred feet long and half as wide. It appeared as though a million cartloads of crushed rock, coarse rock, and earthy matter had been there. The mass was acted upon by loosening percolating waters, by frost, and by sun heat. Two years later, the dumping place was overgrown with grass, gentians, columbines, and other wild bloom.

Frost is a kind of slow, silent dynamite which shatters rock large and small. Ice forming in the cracks and cavities of rocks each winter wedges and forces them apart. Cliffs are wedged off, rock layers split apart, and the earth's stony surface undermined. Deep freezing expands and upheaves rocks far below the surface, allowing the entrance of air and water, which change them into soil. Soil thus is forming beneath soil as well as upon the surface.

Soil in inconceivable quantities is one of the byproducts of a glacier. Possibly one half of all the soil of the earth has been prepared, transported, and spread by glaciers. The glacier grinds rock to powder, which the chemistry of nature changes into plant food or soil. The greater part of the forests, the wildflower gardens, the grassy meadows of the Rocky Mountains and the Sierras, and vast quantities of grass and hay for animals and grain for ourselves in the Mississippi Valley, are grown in rich glacial drifts or deposits.

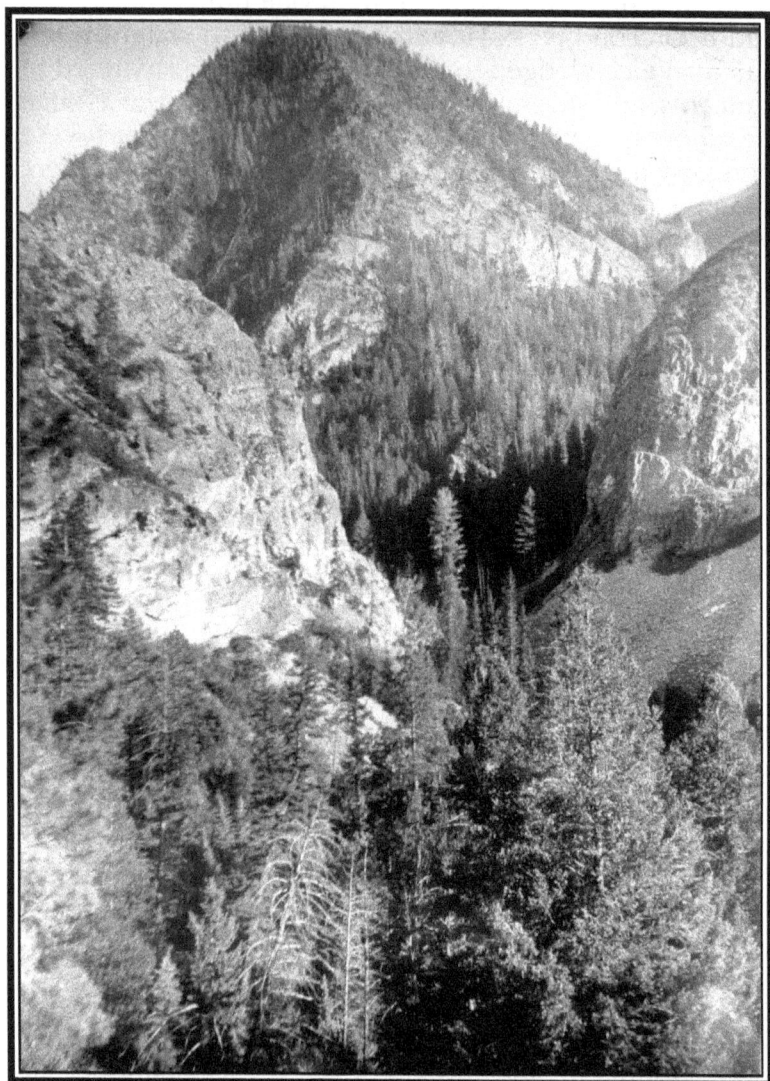

Soil creation in the San Juan Mountains, Colorado.

The vast quantity of material which the glaciers transport in the form of broken rock, boulders, cobblestones, pebbles, and rock flour now reposes in countless moraines, some of which have an area of thousands of square miles. Glaciers thus have been one of the greatest of eroding agents, one of the greatest manufacturers of that fundamental resource—soil.

Roots are a hidden but an important factor in soil production. Each tree has hundreds of rootlets; shrubs, flowers, and grasses each have many roots and rootlets. In an acre, there are countless thousands of these penetrating the earth. A rootlet frequently forces itself into a tiny, almost invisibly crack in the rock. It acts as a wedge through growth expansion. Dissolving water finds its way in. Vegetable tissues, roots, leaves, grass, and logs, when deposited in water, result in the formation of marsh gas, carbon dioxide, and other gases. Deeply penetrating tree roots allow the water and gases to follow down and penetrate as deeply as they. Their roots also carry dissolving acids. These hasten rock decay. Thus, beneath the surface of every acre, there may be thousands of wedges of roots and the action of water, ice, and chemicals shattering rock, which is finally dissolved and reduced to soil.

Trees are sometimes planted in coral rock by drilling a hole, much as if digging in ordinary soil. The lime of the coral is so readily soluble by the water and acids from the roots of the tree that soil forms rapidly, and trees thus planted have been known to flourish and bear fruit within three years; yellow oranges from white coral!

Lichens, while look like rock stain or tinted wall paper, are also an extensive soil producer. Lichens cover millions of square miles of rocks and dissolve with their acids.

Earthworms move and churn up the soil, many million tons each year being accredited to them.

Heavy rains in tropical countries supply moisture so freely that they help produce rock disintegration without the frost which works in the drier regions of the North.

Erosion seizes much surface soil and carries it away. This loss of soil is constant, and in places soil is occasionally removed faster than the new soil is manufactured, thus exposing barren rocks. This erosion is most rapid on steep slopes, in windswept areas, and in places temporarily barren through the destruction of roots which anchored the soil or of the covering which protected it. Wind and water move and remove soil, sometimes transporting it long distances, though often the move is a short one.

It was a day of landslides on Mount Coxcomb! This peak is not one of the "eternal hills" but a crumbling, tumbling, transient mountain. After months of drought, excessive rains had loosened and unsettled the earth and rock of this permanent looking structure. Cliffs tumbled and the sides of the mountain crawled. After counting the crash and echoing roar of forty-three falling cliffs, I ceased counting. A number of the slides of that weird day went to the bottom of the slopes. An absence of a year shows the visitor many changes in the skyline here.

Near my mountain cabin, a ragged-edged meadow has been born where once a lake rippled and sparkled in the sun and lived its liquid day. An intermittent stream coming down to the lake bed from a barren rock slope gradually filled it with sediment; it was slowly covered by a growth of vegetation, was buried and forgotten. During each storm, the silt-laden water drops its load on the edge of the meadow, building a delta and covering naked rock. Grains of sand, sediment, fallen leaves, withered plants, and insect remains are brought together by storms and combined by water into soil. Seeds are brought to new soil beds by birds, wind, and water.

A volcano often covers thousands of square miles with an inch or so of ashen rock dust. This is almost instantly changed into soil. In places, fresh lava from the Hawaiian volcanoes is covered with vegetation at the end of three years. Presto! chemical change, and lava becomes bananas! Extensive farmlands in Nebraska and Kansas are of volcanic dust. Heat, rain, and sunshine make this dusty flour.

Soil production is ever going on at increasing speed, and the wash and wear of wind and water never cease but in places are working with increased rapidity. The result of all this is that a great quantity of soil is washed and blown into the sea and lost. much removed soil is overspread upon existing soil, so that the depth over most areas is deepening and the quantity of soil is increasing.

An eroding mountainside.

A campfire at Two Medicine Lake, Glacier National Park. Photographer unknown.

Ups and Downs of the Grand Canyon

The overturning of my boat in the upper end of the Grand Canyon caused me to rise from the waters heavily laden. In less than a minute, my clothes picked up fifteen to twenty pounds of sand, fine rock flour, mica, marble, and iron, and suddenly developed a strange stiffening overload of armor that impeded every move.

There are two pounds of sediment in each gallon of water of the Colorado River. Multiply this overload of sediment a few million times for each gallon of water that has flowed through the river bed down through the ages, and one has a small conception of the vast quantities of material that have been eroded off the 15,000 square miles of the Colorado plateau and washed away by the caving in of the walls.

The Colorado River has made a distinct showing on the globe. It flows through a series of twenty vast canyons that have a total length of about one thousand miles. Rising on the western slope of the Rocky Mountains, snow-born streams bring to it contributions from high peaks and mountain valleys, and the eroded material from several states. Added to this is windblown dust and sand from desert plateaus. Continual carving and caving of the walls compel the river to spend most of its time and energy in breaking up this debris and carrying it forward to the sea—building up the delta of the Colorado. There are long stretches of quiet water through gentle valleys and low hills, and numerous turbulent currents between steep rocky walls where wild, foaming rapids spend their energy against the river bed. Through its thousand miles of canyons it has a fall of more than four thousand feet, unevenly divided.

Erosion is the artistic agency which causes old landscapes to melt away and new ones to advance upon the scene. This planet is old, and all the material in it, all that is now on the surface and much that is miles beneath the

From the north side of Greenland Point, looking into Marble Canyon.

surface, has been through countless changes. Starting as a lifeless mass of solid rock, without a particle of soil, the earth has been given its present surface by time and the elements; erosion and decay have been lowering and leveling the heights, cutting valleys and canyon walls, washing the loosened material upon the lowlands, and building up new shorelines in ten thousand deltas out into the sea. What is deposited today will be eroded tomorrow. In how many times and places has each particle been deposited, vulcanized, stratified, cut to pieces, and broken up, then transported by wind, water, and ice to form new landscapes in the sun? Through erosion's agent—running water—the earth flows from one form into another.

The triumph of geology is in demonstrating that the forces which we see in operation today are sufficient—given time enough—to explain the sculpturing of the earth's surface.

The Colorado River has been working overtime for untold ages in producing that masterpiece of erosion, the Grand Canyon, which might well be called the greatest wonder in the world. Its immensity alone would be sufficient to attract the attention of the world. Combine with its vastness the marvelous and intricate sculpturing of its walls, the display of all of nature's colors in orderly array, and the appealing geological story revealed in the rocks of sea-born origin, and you cannot find its equal or any to compare with it the world over.

The Grand Canyon is the eighteenth of the series of canyons of the Colorado, counting downstream. This stupendous canyon, one mile deep and twelve miles wide, has been cut by the scratching of the sand and gravel dragged along by the river during an inconceivable period of time. Written in the two hundred miles of its exposed rock strata is a wonderful story of the past. The world-making processes are graphically exhibited in a stupendous panorama of prehistoric horizons. In no other place in the world are all the geological ages displayed in such a magnificent and complete array.

The Grand Canyon is a monument to Nature—to her achievements in uplift and subsidence, rock-making and breaking, sedimentation and erosion. Here are the original Archean rocks—granite and gneiss; sedimentary rocks as formed and as changed through heat, pressure, and other metamorphosing forces; lava new and old, by itself and forced between strata of other rocks; displays of uplift and subsidence; faults, displacement, crystallization, and color; rocks folded, crumpled, deformed, and tilted; a thousand illustrations of sedimentation and the ever-changing story of erosion.

The Grand Canyon plateau has had seven separate lives or existences; three times it has been sea bottom and the ocean covered the region it now occupies; during its three separate submergences, not less than 30,000 feet of rock layers were deposited upon it, 26,000 of which were eroded away. Two of these ancient deposits were completely eroded off, and at present there are 4,000 feet of sedimentary deposits remaining upon it, through which the Colorado River has cut its river bed, and it has cut still another 1,000 feet into the original granite of the earth.

A tourist today picks up sea shells on the rim of the canyon, now 6,000 feet above sea level. This plateau surface of the canyon was formed beneath the sea. In its ups and downs, the Grand Canyon country has been 15,000 feet or more above sea level; as sea bottom, it has been 15,000 feet or more below the waves. Each time it was beneath the sea, numerous strata were deposited upon it. Each time it has been land, the surface has been leveled off and worn down by wind and water.

The outer surface of the earth was never long, if ever, in repose. Parts of the first land areas sank beneath the level of the sea, while other sea-covered areas, commonly those close to the original land, where uplifted into the sunlight. The surfaces raised and lowered have been both the original granite and surfaces of sandstone, limestone, and other rock layers formed upon the original granite.

The Grand Canyon plateau shows a mile of sedimentary rock, layer upon layer, in the original order in which the deposits were laid down. These sedimentary rocks, as well as their embedded fossils, tell definite information. A microscopic examination reveals whether they were deposited in fresh or in salt water; whether the sediments were brought by wind or by water. Cross-bedded sandstone often is of windblown desert sand. Sediments deposited in arid regions commonly form red rocks—the red beds of geology. Most limestone is formed of the minute remains of marine animals. Rock salt and gypsum commonly are the dried-up remains of lakes or sea arms and are the result of evaporation exceeding rainfall. From a thousand ages of the past, the plants and animals of ages indefinitely removed are brought to light in rocks formed of sands, muds, and limes laid down as sediments. Uplifts have raised many of these deposits, and erosion has uncovered and exposed the fossil records made long ago.

Near the El Tovar Hotel, the Colorado River flows in a narrow river bed 5,000 feet below the rim of an age-old canyon. Between the rim and the river are 4,000 feet of sedimentary rock layers, limestone, sandstone, and shale, and a thousand feet of the ancient Archean rock—the original granite and gneiss of the earth's surface. The history of the cutting of the Grand Canyon covers an inconceivable period and takes us back to the earliest history of time. From the enduring records told in rocks, we can rebuild the past.

This Continent, if mapped a few million years ago, would have been very different in its land and water areas. The Gulf of California reached far up into the north, extending into Nevada and covering the areas now occupied by the Mohave and the Colorado deserts. The Gulf of Mexico was vastly larger, extending up into the Mississippi Valley toward the Great Lakes, and completely covering Florida. Slight uplifts along an inland sea brought new lands into existence. And every bit of land above the waves was material to be battled with by wind and water, frost and ice, and all the elements helping to wear it away.

If we can imagine the Grand Canyon country when it arose from the primeval sea, we can see its surface being slowly changed in form and appearance, until the original granite was worn down to a level plain and the eroded material was reposing in the shallow waters of the inland sea.

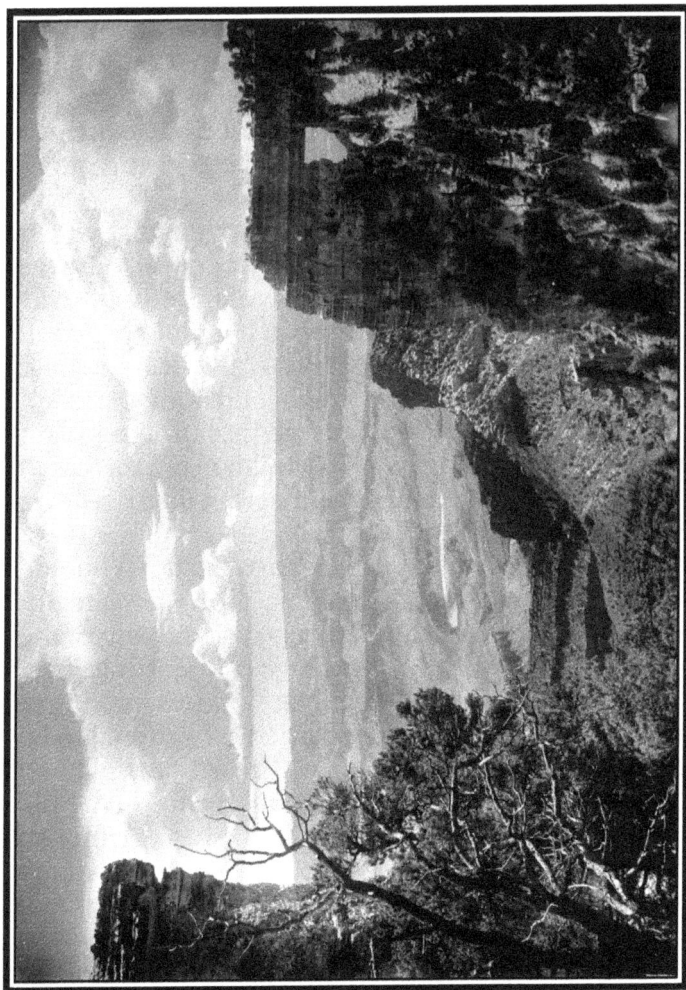

"When one beholds this marvelous, sculptured canyon for the first time, other scenes and places are forgotten—there is no comparison."

The Archean rocks slowly subsided. During the ages that the inland sea regained possession, erosion was going on upon all the surrounding land areas, and 12,000 feet of sediments were laid down under water. These deposits formed layer upon layer of limestone, sandstone, and shale. At this time, so fossils found in the remnants of the Algonkian strata show, the earliest forms of life were developing. During this epoch, which lasted an interminable time, the region was the scene of volcanic disturbances, as evidence in dykes and chimneys of lava shows. These disturbances shattered, cracked, and seamed the strata.

Then came a mountain-making movement which uplifted these 12,000 feet of rock layers into the sky. It was accompanied by faulting and tilting. While the uplift was occurring, erosion was working away at the tilted material, and eventually about 11,500 feet of the 12,000 feet that had been deposited were worn away. During the long period of erosion, a mountain range was reduced to a rolling plain again, and only small remnants of this brilliant red tilted and faulted Algonkian strata remained as insets in the base level surface of the old original granite and gneiss foundation of the earth. There were probably rivers then, as now, but the Colorado River had not yet taken its place on the plateau.

Another long period of submergence and sinking of the plain followed, allowing the incoming of the Tonto Sea, which worked upon the surface, still further wearing it away. From nearby peaks erosion carried off material to be deposited upon the submerged plain. The sea was salt and shallow and it is very likely that the submergence of the old Archean granite and Algonkian plain was going on simultaneously with the depositing of mud and sand of the Cambrian epoch. Then a rest, and these deposits were cemented into sandstone and shale—the strata now seen resting on the remnants of the Algonkian strata and in places upon the original granite. Two hundred feet of deep buff and greenish-gray strata of the Cambrian epoch show strikingly in the lower walls of the canyon, beneath the Tonto platform. They are much duller in color than either the brilliant crimson of the Algonkian below or the red of the Carboniferous strata above. The Tonto platform forms a distinct, wide terrace, owing to the capping of these more resisting materials.

The records of the Ordovician, Silurian, and Devonian epochs are almost a blank, and it is probable either that the sea retreated for a time and no deposits were laid down, or that there was a slight uplift and sediments laid down were exposed to erosion. Only slight evidence of Devonian deposits has been found in the region.

During the Carboniferous epoch two thousand to three thousand feet of deposits were laid down. These formed the red-wall limestone, cross-bedded sandstone, and cherty limestone now exposed by the cutting of the Colorado River. The strata of red-wall limestone are the most striking in the canyon, forming the greater height of the present wall. There was no coal formed in this region during the Carboniferous epoch, the arid conditions of the climate and the lack of vegetation determining the materials laid down in the water. The purity of the limestone is evidence of the clearness of the waters in which the deposits were laid down and the presence of large cup corals shows that the sea was warm. The sea was probably shallow for a time, as alternating deposits of sand and mud were laid down, forming the shales and sandstone. On top of the Carboniferous, continuous deposits were probably forming during the Permian, Triassic, Jurassic, and Cretaceous epochs, until two miles of rocks accumulated.

A great mountain movement occurred in early Eocene times, accompanied by violent disturbances of the earth's crust and many changes in the plateau country. The Cretaceous rocks and underlying strata were uplifted and subjected to the fierce attack of erosion. The records are not complete, but it is assumed that a great warping occurred; some of the plateau sank and other portions arose, forming an inland sea. Vast deposits occurred again, their weight sinking the plateau deeper and deeper. Alternating uplifts and subsidences without record were at work during a great extent of time.

During the continued slow process of uplift, the Eocene lake was drained and 15,000 feet of deposit that had accumulated since Carboniferous times rose above the sea.

During eons of erosion, this 15,000 feet of rock strata was entirely washed away. Only slight traces of the Permian, Triassic, Jurassic, and Cretaceous epochs are found in detached underlying strata of the Grand Canyon. And the subsequent deposits of Tertiary times were entirely removed upon the Grand Canyon rim. Peripherally some distance back, north and south from the rim, the eroded and broken strata of rock layers subsequent to the Carboniferous epoch are well defined.

The Colorado River was probably the outlet of the freshwater Eocene lake that disappeared during uplift. During this time, the river flowed in an easy going shallow channel, in a comparatively level plateau. As continued uplift gave it impetus it cut its channel deeper, and as erosion was hastened, the debris-laden water sawed deeper as it slid over the surface of its inclined plateau. There were pauses in the uplift and a slowing down in erosion, but with each new steepening of the grade, the opportunities for erosion were increased. The climate was probably one of increased precipitation, and erosion of the entire plateau was undoubtedly going on at a rapid pace.

Enormous periods of time are represented in the fact that all the later geological epochs are missing along the rim of the Grand Canyon. The epochs had their existence, as rocks deposited elsewhere show, but in this region these thick layers, slowly deposited, were again slowly eroded. Sections of the adjacent plateaus show more recent periods of geology, and from the remnants of these strata, geologists restore the formations which once rested upon the Grand Canyon rim. Not all the strata are complete in one place, but where in horizontal position, or preserved by the cappings of volcanic material, they have endured through periods of subsidence, uplift, and erosion. The remnants give the most complete and logical abstract of all the geologic epochs anywhere to be found.

It may have been fifty million years since the original granite of the Grand Canyon region first rose above the surface of the sea. And it may have been twice as long. Slowness is the keynote to the marvelous story of geology. Slow upheaval and slow settling; slow erosion and slow

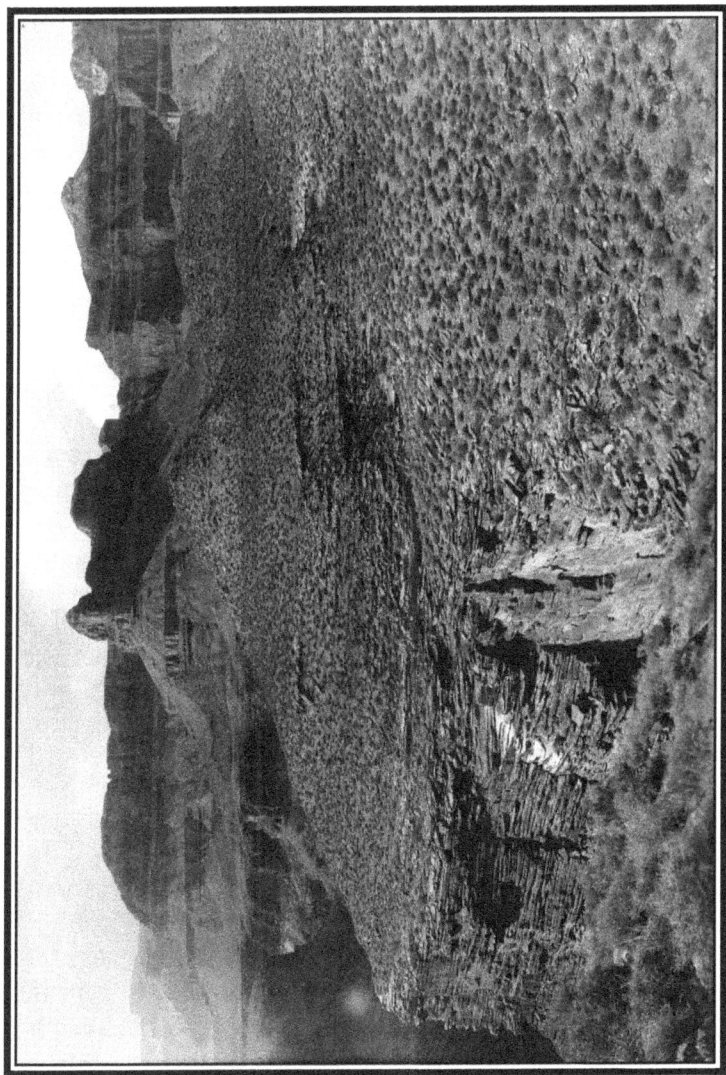

A view of the Grand Canyon on Hermit Rim Road.

forming of rock layers. One hundred years or more for each foot of uplift might be called average; an entire century for a few inches of subsidence would be normal.

A few small land areas are being thickened by additions of sediment or dust, volcanic ash or lava, but most of the surface of the earth is being lowered by erosion, which goes steadily on, much less than an inch of a century. Sea bottom over most of its area is being built up by hundreds of millions of tons of rock sediment annually poured in by rivers and the dusts contributed by winds and volcanoes. Sedimentary rocks are deposited inch by inch, in some cases less than a foot a century; in other cases or places, the rate of deposit might be trebled. This material is combined into many kinds of rocks. A little more of this mineral or that chemical, a trifle more or less heat or pressure, and the rock produced is distinctive. Limestone intensely heated becomes marble.

Layers having a thickness of approximately 15,000 feet have been washed off the present surface of the plateau. And this surface still is being lowered by erosion, and possibly, too, it is also being uplifted by upheaval. But if conditions were to go on very much as now for, say, one million years, in this time the walls of the canyon would widen and flatten out into moderate slopes. The removal of 4,000 feet more of sedimentary rocks would uncover the original granite. Sometime this 4,000 feet may be completely removed; sometime this old granite may again slowly sink, give the plateau region over to the sea, while, for the fourth time, as sea bottom, it receives another covering of sedimentary rocks.

Of all the works of nature of which we have record, the Grand Canyon has been the slowest and the most eloquent. Three times submerged beneath the sea, four times uplifted into the sky. What time and transformation are represented in the strata laid down by the slow processes of the elements! Then slowly, tediously uplifted into a mountainous plateau, to be slowly planed down again by running water, ceaseless winds, rains, frost, and extremes of temperature; more accumulations of sediments through thousands of years, and another prolonged battle with the elements. Two complete deposits entirely worn away, washed out of the landscape, and 4,000 feet now in process of being removed to other scenes.

Four prominent factors have been favorable to the production of this architectural wonder: the elevation of the plateau, the horizontal stratification, the varying degrees of softness and firmness of the rock layers, and the aridity of the climate.

A narrow ditch will widen by the gradual caving in of its walls. The sharpness of these canyon walls is due to the peculiar climatic conditions, the short rainy seasons and long periods of drought. The lack of vegetation is everywhere evident, giving erosion little resistance and displaying to the best advantage the relief work of rock sculpturing. Spasmodic rainfall and cloudbursts, followed by long dry seasons, turn dry rivers into torrents, with an eroding and carving power of magnitude. And during dry periods, the ever-busy wind takes up the work. It is likely that, through the ages, winds have ever been a factor in erosion, in shaping the topography of the earth. Much of the cross-bedded sandstone which abounds in the walls of the Grand Canyon had its dry day with sifting, shifting winds; in dunes and drifts, it marched and dashed beneath cloudy screens of dust and the sweeping roar of storm.

In the desert of the Grand Canyon plateau, one may see the ancient cross-bedded sandstone now on the surface being eroded away by the wind. The closest call I ever had of becoming a desert fossil was on the Painted Desert, not many miles from the Grand Canyon. After thirty-six choking, thirsty hours in a sleeping bag, I crawled out and found myself almost sealed in one end of a new-formed sand dune that was more than two hundred feet long, forty feet wide, and three to eight feet deep. All in one storm it formed, by a rocky outcrop, and much of it must have been swept more than a mile. The air was so full of rock powder during the storm that for hours I was threatened with strangulation. Horses, men, and wild animals occasionally are buried alive in the dust drifts of these desert storms. In this storm, as far as the eye could reach, the low-lying desert topography in every direction changed.

Geological records show that the desert has ever had large land holdings—about one fifth of the solid surface of the earth. Many millions of years ago, this same sand now being broken from the surface sandstone was being blown and drifted about on the prehistoric desert of the Grand Canyon region. Erosion, rebuilding, submergence, stratification, uplift, and slow erosion again in which the wind has had its part is the story of the sculpturing of the Grand Canyon walls.

The making of the Grand Canyon has not ceased. Running water, the smooth-edged winds, and the silent frost—the age-old tools of the elements—never wear out. They work with every varying device known to nature, changing with the material which they meet. Even the gentle raindrop grapples eagerly with mountains of solid rock, the softer materials receiving the deeper impression, the harder strata resisting longer.

When one beholds this marvelous sculptured canyon for the first time, other scenes and places are forgotten— there is no comparison, for there is no object in nature on the same scale of grandeur. And with a longer acquaintance comes a desire for a thousandfold capacity of greater feeling and comprehension.

I have boated in many of the canyons of the Colorado and have camped and tramped along their rims. I have looked down into broken depths when they were filled with mists; when colored clouds hung over them, at sunrise and sunset, I have watched the mysterious changing lights transform the age-old strata with the intenseness of rainbow hues.

While the extent of the river through the Grand Canyon plateau is 217 miles, this does not suggest the ins and outs of the canyon walls around out-jutting peninsula-like promontories and into deep recesses, all carrying their orderly banding of red, brown, gray, buff, and crimson. It is like looking down upon an inverted, hollow mountain range, with ridges, spurs, plateaus, cliffs, shattered pinnacles, broad platforms, detached peaks and buttes. Stratified color in a magnificent assortment; walls of brown and red, deep layers of gray, yellow, grayish brown, and green—all combine to produce a landscape in form and color unrivaled, and so harmoniously balanced as to height and breadth as to suggest perfect architectural achievement.

Lying in the still, clear air of the western desert it seems finished, complete and unchanging. But even in a torrential rain, the serrated cliffs and pinnacles and sloping taluses stand unmoved. Rocks may roll and cliffs may fall while streams wash on, but standing out in the heart of the canyon, four thousand to seven thousand feet above the river bed, the massive buttes and terraces are all one sees. The scale is too immense to distinguish detail. It is the whole that makes it all, like the thousand whitecaps on the lapping waves that break and mingle to form again; the myriad leaves of a forest that lift and stir the sun and air. And then the alchemy of color that binds it all in one, that underlies the mystery and the magic of the sculpturing, bursting from the hidden recesses of the deepest strata, edging the serrated cliffs, and outlining the buttes and mesas abutting the wider sections of the canyon. Even the air is filled with mystical lights of blue, violet, and lavender, or gold and yellow, according to the time of day, as though caught up in light rays streaming through the canyon walls.

Color is the impressive characteristic of the Grand Canyon. Desert chemistry develops colors of the most brilliant dye. The absence of vegetation allows the fire-toned and time-colored rocks to be seen in all their brilliant richness.

This arid land makes the most lavish and artistic displays of deep rich colors. Many a sunset I have seen in which the broken horizon clouds were of melting opals or melting gold. Dawn often came up, not like thunder, but lurid, as though below the horizon were only volcanic flame and smoke. Under the slanting evening light the heights, spacings, and outlines of the mountain walls show with bold distinctness and their colors with all their freshness. The canyons fill with solid purple while the white up-shooting last rays give the upstanding peaks their blackest silhouettes and sharpest outlines.

Geology is intensely filled with information appealing to the imagination and, as well, vital to our welfare: the story of coal, the romance of soil, the history of minerals, the combining and recombining of materials of this old earth. For its cycles of change through unknown periods of time, its stories revealed in rocks, fossils, and erosion, its existence through uplifts and subsidence, the greatest production in nature yet revealed is the Grand Canyon of the Colorado.

Photograph by Kolb Brothers, Grand Canyon, Arizona.

The Rock Cycle

Rocks are everywhere about us, rocks solidly in place— part of the bedrock of the earth, and boulders in stream beds and glacial deposits that have traveled far from home. A boulder could tell an interesting rock biography—what it is, where from, how formed, and what were its possible cycles before.

Erratic granite boulders brought by the ice from far away Hudson Bay tell of the glacial epoch, during which the Missouri River was forced southward nearly fifty miles by a great wall of ice to cut its present valley, leaving its old valley to be occupied by Milk River, a much smaller stream.

One vacation, many years ago, I drifted a thousand miles down the Missouri River in a boat, camping at night along the bank. I had several happy days with a geologist who had come exploring up the river. As we sat on the bank one noon, he reached down, picked up a stone, looked at it, broke it, examined the break with a magnifying glass, and then said, "Although the lower channel of the river is comparatively recent—cut perhaps at the close of the last Ice Age—here is a stone at least four million years of age, and perhaps twice that. It was formed in an ancient sea, made of consolidations of worn-out structures of minute sea life and deposited in a stratum." He pictured the scenes and its neighbors at the time it was forming. Growing in the sea nearby at that time were numbers of now extinct plants. Around this forming stratum swam numbers of uncouth sea fishes and sea reptiles of gigantic size, but these became extinct ages ago and are known only through fossils.

We sat long discussing the biography of this stone, how it came to be where it was and what it was. And it was still susceptible to endless combinations—might become a part of other stones, might cement sand into sandstone, might sometime be roasted and become fiery limestone again— even after numerous other combinations—or richness for cold, barren soil.

Geology is a history of the earth as read in the rocks; and from this history we learn of the composition and structure of rocks—how they were made and how modified. Everywhere, in fields and plains, along river beds and the seashore, in mountain cliffs and canyons and upon the desert, we may see rocks of one kind or another.

Granite is one of the First Families of rocks. It dates back to the beginning of the earth, and from it many other rocks have taken their materials. A piece of granite may have had a part in making sandstone, shales, clays, schist, or lava. The original granite of the earth has gone through fire and water in creating the varying forms and types of rocks that now require a corps of scientists to identify, differentiate, and classify. Flood, earthquake, and volcano have kept the surface of the earth in a state of quandary as to where and what it was. Heat and pressure below the surface have been silently, invisibly at work remaking the granite, composing new materials to display to the all-seeing stars. The geologists say, "Given time enough, nothing is more changeable than rocks."

Granite is of composite structure. The numerous minerals within the earth have combined with the original substance to produce a wide variety of types. Some of the minerals conspicuous in granitic rocks are feldspar, hornblende, mica, and quartz, giving colors ranging through light and dark gray to pink. Granite crumpled, mashed, and broken up under intense pressure is known as gneiss, the commonest and most evident formation on mountain summits where erosion has worn the earth's surface down to the core. Granite, being the oldest rock, has been most metamorphosed, gone through most changes. Intense heat within the interior of the earth has thrust up the granite as igneous rock, intensely changed through heat and the action of water and steam. All rocks may be subjected to these metamorphic processes, either within the heart of the earth or upon the surface. Sedimentary rocks are susceptible to change; quartzite is a type of metamorphosed sandstone. Limestone, when metamorphosed recrystalizes and forms marble; shales, under heat and pressure, develop a series of rocks which approach in character the igneous rocks from which they were originally developed, and may become

Gunsight Mountain, Glacier National Park.
Photograph by Fred Kiser.

schist.

The mountain ranges of all the continents have uplifted upon their broken skylines the original granite, metamorphosed granite and igneous rocks in seams, dykes, and fractures; and around their bases—on the surface or still deeply buried—are the stratified rocks of sandstone, limestone, shale, and a multitude of combinations of all these materials. Minerals and chemicals and water have combined with the original granite to produce entirely different rocks under varying conditions, and the degree of breaking up and circumstances of remodeling have resulted in the mineral and rock resources of the earth. Through these glorifying processes, deposits of gold, silver, lead, and copper were produced. Marble, coal, chalk, gypsum, and salt are some of the byproducts of rock making that are now being mined for the benefit of mankind. Vegetable and organic matter have had their part in compiling rocks. The Great Pyramid in Egypt is made of limestone formed of shells of tiny sea animals. In Alabama, there are chalk beds one thousand feet thick—chiefly the accumulation of the skeletons of minute plants and animals.

Upon the original granite, the forces of erosion, transportation, and eruption placed layer upon layer of varying kinds of sediments. Lime and other minerals cemented the sediments deposited in shallow waters of inland seas, lakes, and river deltas into limestone, sandstone, and shale. Eroded granite and volcanic rocks with their minerals and chemicals mixed with water were laid down in the sea to form a very different rock than either.

In quiet waters, as in the ocean depths, these strata were nearly horizontal and were parallel and of even thickness. Where changing currents swept the loose and soft materials about, the layers were more or less irregular in thickness and deposition. They may have been cross-bedded and laid down at considerable inclination. If the streams were bringing chiefly sand to the ocean or sea bottom, then the rocks formed would be sandstones. During one great geological period sandy limestone and sandstone a thousand feet in thickness was laid down in what is now the Mississippi River valley. This is now being exposed and can be strikingly seen at McGregor, Iowa, along the Palisades of the Mississippi. If the material deposited was chiefly mud or clay, layers would be laid down, which, in time, might be compressed into shale. Far out from shore, where the waters were still and clear, there might be formed layers of limey material, either by chemical action and reaction or by the work of organisms.

The sedimentary, stratified rocks are the newest and contain the most recent record of the earth's history. These sedimentary deposits are estimated to be forty miles deep. They are formed of deposits of varying materials—broken rocks, soil compounded with minerals, chemicals, organic and vegetable matter, under the influence of water, heat, and pressure. Of course, these layers are not now found all in one place or in regular order. They have been classified, and the rock layers the world round arranged into one system in chronological order. Their benchmark names have been suggested, commended, and almost commanded by the material, color, structural peculiarities, and fossil models found in each layer. The finger of time so marked the rocks that they might be said to keep the time records of earth's history. The order of succession or superposition of the beds must necessarily be the chronological order of formation. The lowest beds must be deposited first. The records of the vast changes through the past are read in rocks almost as conclusively as though we had seen them in action. These events are revealed not only in fossils of the plants and animals that left their record in the rocks, but in rocks themselves that are fossils. The fossil plants and animals found in rocks make it possible to compare the chronological order of rocks found in different parts of the earth.

The conditions that prevailed during the geological periods of the past are revealed to us largely through stratified rocks; from the color and character of the deposits and the records of plant and animal life, we are able to tell the climate, the nature of the topography, and many of the dramatic events in the formation of the earth. Though these records are incomplete, they are in many respects conclusive and dependable, and by means of them, we have

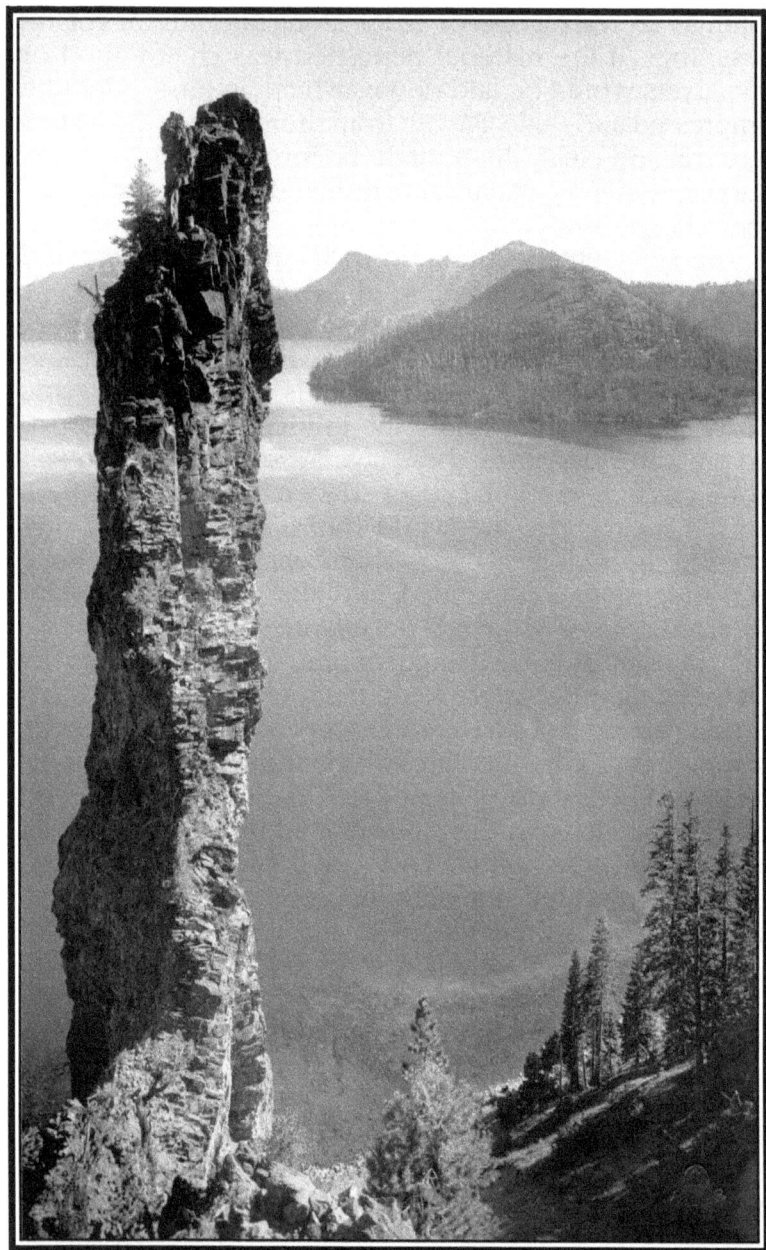

Wizard Island from Devil's Backbone, Crater Lake
National Park, Oregon. Photo by Fred Kiser.

abstract and brief chronicles of millions of years of earth history. The dead of a million and more years ago are brought back to life again and ancient scenes restored.

Climate being a profound factor in determining the character and distribution of plants, if the climatic requirements of a fossil plant be known, this fossil will tell of the climatic condition of the locality where it was growing. Fossil palms are found in Greenland; these reveal the fact that at one time the region was much warmer than now. "Fossil plants," says Asa Gray, "are the thermometers of the ages by which climatic extremes and climate in general through long periods are best measured." A knowledge of fossils—the study of paleontology—gives one a grasp on the life history of living plants, a knowledge that explains WHY living plants have come to be WHERE they are, and also HOW they have come to be WHAT they are. This knowledge enabled the sequoia species to be known and to be named as a fossil before it was discovered and named as a living tree.

During the middle Jurassic epoch, and probably over a long period prior to this time, plants of all kinds appear not only to have been associated, but to have been distributed over the earth, almost uniformly between 63 degrees south latitude and 82 degrees north latitude. Apparently, there was continuous land for thousands of miles north and south, and as various kinds of plants scattered over this area show like growths, and as they are without either growth rings or annual rings, these fossils record that the earth in those times was without zones or seasons. Annual rings first showed during the late Jurassic epoch.

Fossil reports show that the climate of the whole world grew cooler and dryer during the Miocene epoch. In western America, it was arid, very much like the present. In this epoch, copper deposits in Montana and gold and Cripple Creek and in California were forming. There were boundless forests and no end to wild flowers. Along the Pacific, the present oil and coal supply was being pressed and compounded upon a sinking Miocene surface.

Coal beds around the world tell of strata laid down during the Carboniferous period, formed of vegetable matter

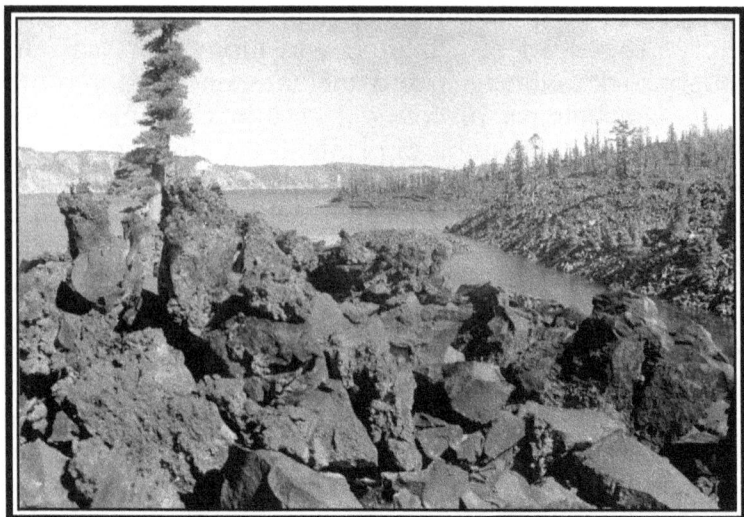

Above: Lava fields of Wizard Island, Crater Lake.
Below: Sunset from Wizard Island.
Photographs by Fred Kiser.

and the minerals which these plants drew from the soil. As plants died, they fell into water and were preserved from decay. Layers of this luxuriant plant material were covered with mud or sand and the processes of coal formation began. The weight of superincumbent rocks compressed sediments into shale and sandstone. The masses of vegetation in the coal swamps were compressed. Water and some of the gases were driven out, and other changes took place, resulting in the development of solid layers of coal where originally existed only pulpy beds of stems and leaves and fruits. Several successive coal swamps exist in the same locality, separated by layers of shale or sandstone.

A coal mine in New Mexico had been abandoned because the vein was lost. It was supposed that where the coal had disappeared the bed had been faulted out of place and eroded away. A geologist visited the deserted mine and found a fossil in the rocks. The fossil was identified by one of the paleontologists at the National Museum as the tooth of an extinct animal, and by means of it, the geological age of the rock was determined. The geologist was then able to say that the coal bed had not been eroded away. The mine was reopened and has been in operation ever since.

Rocks are full of weather reports of the past; they enable us to restore the climate. Gypsum and rocks salt are accumulated only in salt lakes which show arid climate. A salt lake could not exist in a region of normal rainfall, and from the geographical distribution of such salt lake deposits it may be shown that arid conditions have prevailed in each of the continents, not once, but many times.

In the numerous rock strata there are forms and facts enough to visualize the geological times. Rocks enable us to restore the shorelines and extinct drainage systems. It is possible to tell whether the rocks were laid down in the sea or on land, or in some body of water not directly connected with the sea, such as lakes or rivers. If the rock is made up of coarse sands and gravels we know that the deposits were laid down in shallow water where they had not been subjected to erosion for long intervals. Finer sands and gravels are carried farther out from the shore, and the deep oozes of the sea have still different characteristics which tell conclusively their position. Strong currents from different directions leave their records in the ripple marks and in the cross-bedding of sandstone. Sometimes beds of sandstone are separated from beds of shale, showing that the shallow waters which favored the depositing of the sands became deeper by the subsidence of the land and resulted in the depositing of layers of muds and clays. Sometimes there are distinct layers in the sandstone or in the shale. This shows that there were intervals when the deposition of the sediments was interrupted. It may represent a period of luxuriant vegetation on adjacent low-lying lands from which erosion was interrupted. This may be followed by intervals of aridity with increased erosion which is recorded in a renewal of deposits. By plotting a map of the marine rocks of a given geological date it is possible to approximate the extension of the sea over the present land for that given epoch. But for areas which were land in the past and which are now sea, there is no such direct evidence, and it is more difficult to restore the land bridges of the past.

The Great Basin contains numbers of lakes that might be classed as hibernating lakes. These are named Playa lakes and occupy shallow flat basins. They fill with water during the winter or the spring, and after a few weeks, this evaporates, leaving the bed paved with hard, smooth cakes of mineral, hardened and sunbaked mud, utterly destitute of vegetation. In time, the sand blast and evaporation cover spaces with saline crystals that are swept off during the next wind. While these lake beds are drying out, their soft mud makes them practically impassible. Many an ancient animal track is now preserved in rock that, while Playa mud, received the footprint.

Hundreds of saline mineral deposits lie beneath the surface of the Great Basin. Uncountable millions of tons are in deposit. These deposits, the fossil remains of many ancient lakes, have long been commercially worked.

In Nevada are strangely rounded rock deposits which Clarence King called thinolite or tufa. It is a combination of saline lake sediment and mineralized water precipitating from contact in lake shores.

Geological in the master exhibit of the Great Basin—

rocks, volcanic, sedimentary, and metamorphic strata at every angle and of most geological horizons, with no vegetation to conceal their context, their color, or their neighbors. Here people who have never had the slightest interest in rocks become interested in geology, and scientists who ever are crystalline and cold become enthusiastic.

With a microscope, it is possible to tell volcanic material from ordinary windblown material or that which has been carried by the water. Rocks which have been solidified from a molten state and have cooled deep within the earth can be distinguished from those masses which have cooled upon the surface of the earth.

The records of the various Ice Ages, which have been preserved by the rocks, give even sharper and more definite standards of time comparisons than do the fossils themselves. They record the characteristic accumulations made by the glaciers and the formation and disappearance of the icefields. But, again, the fossils in the rocks tell which of several Ice periods is being studied. The rocks are the historians for the upheavals which have occurred through the movements of the earth's crust. These, also, give time measurements and make it possible to arrange the rocks in chronological order and to correlate in one system the rocks in various continents.

The history of mountain ranges is told by rocks. It is quite practicable to give a geological date for their initial upheaval and to determine whether one or many such series of movements have been involved in bringing about the present state of things. The history of plains and plateaus, hills and valleys, lakes and river systems, is recorded in the rocks, and for the later ages of the earth, a great deal may be learned in regard to the successive forms of the land surfaces in the various continents.

Each geological age may be said to have had life forms or models peculiar to that age—just as the first automobile model changed for each succeeding year. An old animal or plant model, as well as a motor model, may long survive and mingle with the newer ones, but its peculiar form, if fossilized, would ever proclaim the time to which it

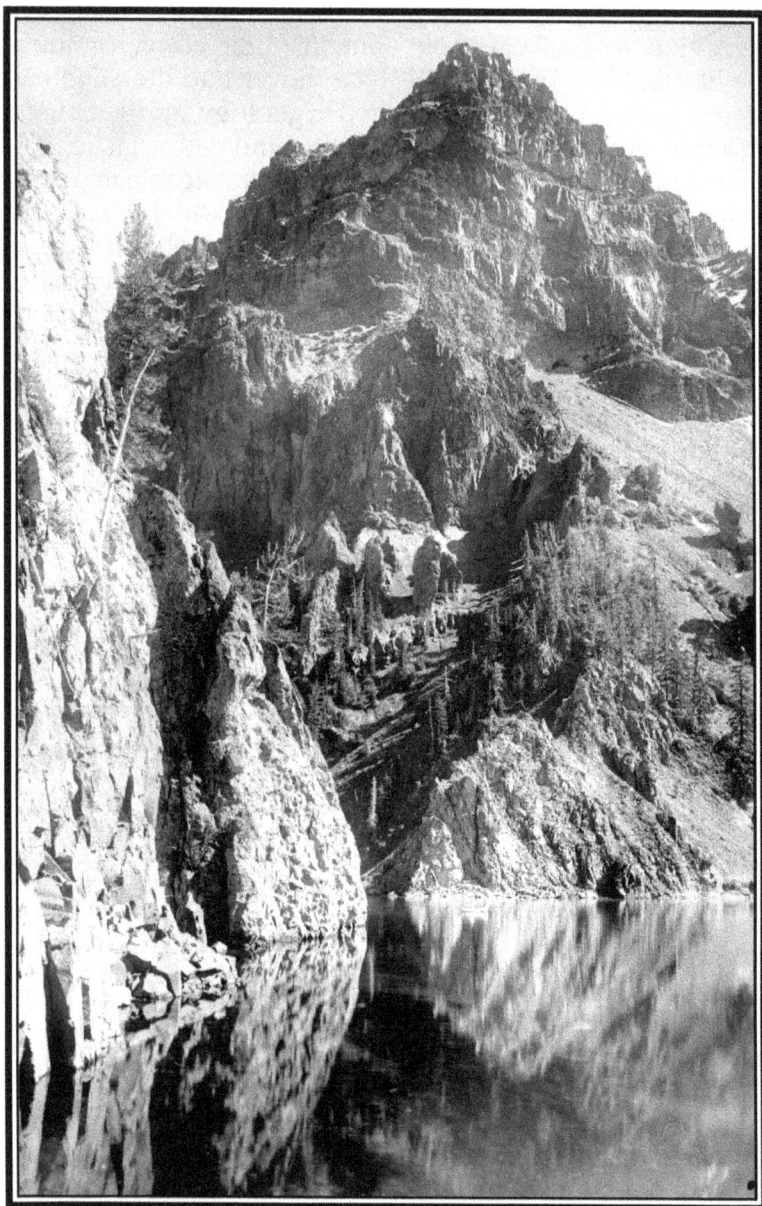

Dutton Cliff, as seen from the Phantom Cliff,
Crater Lake. Photo by Fred Kiser.

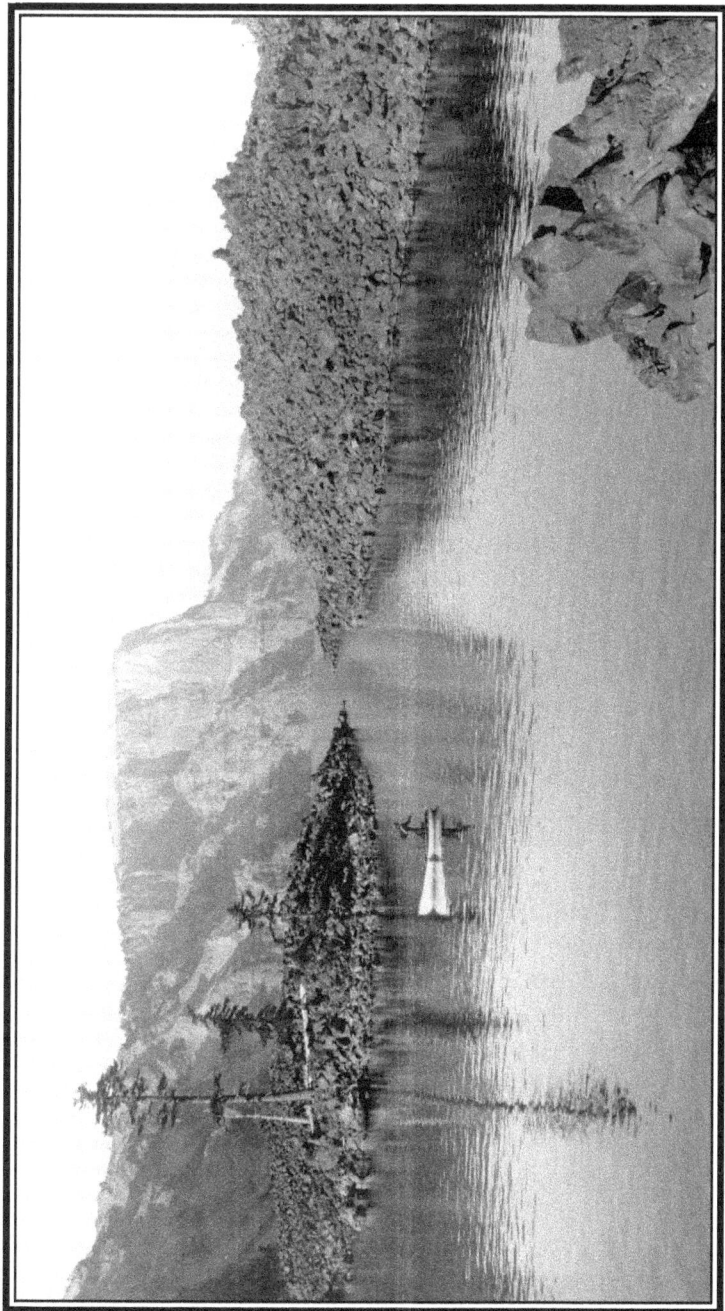

Canoeing past the Lava Fields of Wizard Island. Photo by Fred H. Kiser

belonged. A fossilized automobile would show the form and fashion of the day when it headed the procession, and obsolete chariots of still earlier years would take their place behind. And so it is and was with all life forms. So distinctly formed was the model of each Age, so true do these fossils run to form, that even though a fossil be detached or its stratum overturned, the peculiarities and characteristics of a fossil would carry sufficient information to enable its original place, that is to say, its stratum, to be known. Anyone familiar with the life, the identification marks, of any age, would be able to identify a fossil of that age, no matter where found, as readily as we now determine by its obsolete outline the 1910 model of an automobile.

Life, so fossils show, has existed upon the earth through millions of changing years. The oldest fossils indicate that the beginnings of life were in the waters of the sea. These were trilobites, worms, mollusks, corals, and sponges. Ages went by with life evolving in the sea, and with the land entirely lifeless, without moss, flowers, or trees. Nothing that breathed the air left track or other fossil record beyond and above the shoreline of the sea.

Beginning at the bottom of the estimated forty miles of sedimentary rocks, the first fifteen miles of deposits are without fossils, are barren. Either the earth was lifeless during the periods these sediments were being laid down, or these fossil remains were later destroyed. At any rate, these deposits have not as yet been found. But about the sixteenth mile, that is to say, down twenty-five miles from the present, the first crude forms of life, the first simple fossils, have been discovered. The first life, as recorded in fossils, came abruptly into the scene. Although this life was of crude and simple forms, yet it had the appearance of having evolved from older life and from forms more lowly still; would indicate a long line of ancestry that is a development running through enormous periods of time beyond the first fossil records found. The next layer or mile of strata above the sixteenth, which carries the first forms of life, shows additional forms of life. Ascending through the rock layers many of the old forms and models of life drop out one by one and become extinct, and occasionally a new and better model comes into the scene. Life in each age shows improvement over the life in the preceding age. Thus, each geological rock layer or horizon has a brand or trademark and carries its own special exhibit of fossil life.

As William Smith, father of English geology, said of fossils, "They are to the naturalist as coins to the antiquary; they are the antiquities of the earth, and very distinctly show its gradual, regular formation and various changes of its inhabitants in the watery elements."

Rocks hold the fossil records of many prehistoric horizons. The stone tools of man are considered fossils. Amber ages old often carries complete models and forms of insects, flowers, pollen, flies, seeds, and fungi that became embedded.

A fossil is nature's way of keeping records and writing briefly of the life of other days; it is the remains of past plant or animal life; a track, impression, or record of its former presence. Sometimes this record is a plaster cast, a mold or model of plant or animal. In most cases, the animal or plant fossilized was completely buried or covered in a short time, in sand or mud, which turned to stone. Sometimes the plant or animal was slowly removed by chemical action and a stony deposit left in its place. Occasionally, some plant or animal or bird of today is being deposited where it will form a fossil that will be found and tell a story in the ages yet to be.

Volcanic fires and forces were wildly, vastly active over the earth at widely separated intervals during ages past, deeply overlaying many regions with molten rock. In the Miocene times, Vesuvius first registered in a horizon and began to build and to be a fiery figure in many horizons before Pompeii. Two hundred thousand square miles of Oregon, including the now famous ashen formed John Day fossil fields, were covered half a mile deep with lava. Ashes and flames were pouring from young volcanoes—Shasta, Mount Mazama, Mount Hood, and Mount Rainier—and scores of fire mountains elsewhere over the earth. Their furnaces were flooding many lands with lava. Crater Lake, Oregon, has a story that amounts to a transformation scene. Here formerly stood Mount Mazama, a stupendous volcano

Crater Lake National Park, by H. T. Cowling.

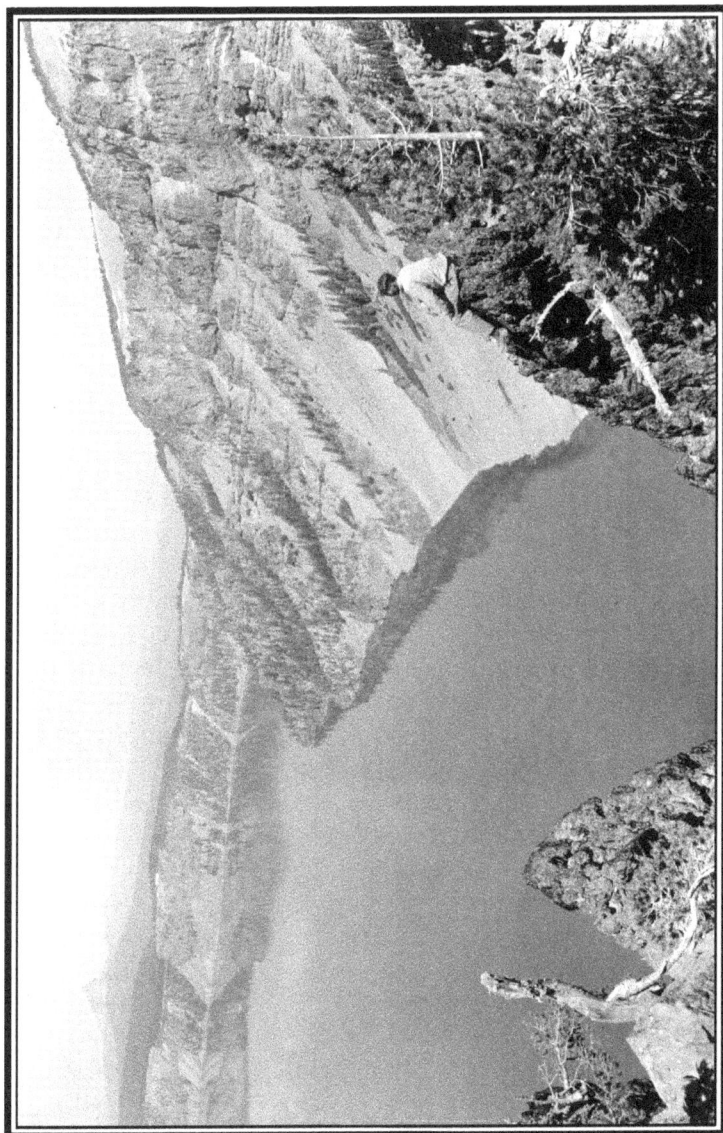

Crater Lake National Park. Photo by H. T. Cowling.

with glaciated sides and a throat more than fourteen thousand feet above the sea. Apparently, there was a volcanic explosion which literally blew off and to pieces the upper six thousand feet of the mountain. The old crater, about six miles in diameter, at once became a lake basin and now is deeply filled with clear water. Its entire appearance is so strange as to suggest that its place was in the landscape of the moon.

Buried beneath volcanic showers are many of the records of the past. The twelve petrified forests on Amethyst Mountain in Yellowstone, presented in agate and opal, are an enduring biography of forest life some thousands of years ago. The soil in which the forest grew, the age these trees attained, the growth of successive summers, and the vegetation flourishing with them are some of the illuminating facts revealed by the cutting of the Lamar River through two thousand feet of volcanic rock. Twelve separate forests, one above the other, are exhibited in the wall of a canyon with a living forest waving on the summit. So perfect is the transformation by mineralized water, the annual rings of the sequoia may be counted in all the brightened beauty of agate. Leaves, twigs, cones, and bark have been sealed in this permanent exhibit. Between the periods of volcanic eruption covering each full-grown forest were intervals of soil making; then lava brought forth another green and blossoming horizon. And over all, the Ice Age left its record. Each layer of these twelve-forest-deep rock strata has several species of trees—pines, dogwoods, redwoods, spruces, and cottonwoods. One of the most interesting of these specimens is the sequoia-redwood. Because of the resinous character of the woods of these coniferous trees they were often preserved to a remarkable degree of perfection. They were not very different from the redwoods of today.

It has been aptly said that "Everything in nature is engaged in writing its own history." Rocks are our geological library. In rocks, the plants and animals of the past have left a pictured record of themselves, of their activities, their habits and customs; tracks fresh, dim, and strange; bones of pygmies and ponderous ones; these are

Petrified trees in Specimen Ridge Fossil Forest,
Yellowstone National Park.

the key that unlock the secrets stored in rocks and enable the great story of geology to be intelligently read. Through our knowledge of fossils preserved in stone, figures and shadows shift and move across the strange landscapes and old scenes of the past.

Although, as a boy, I worked much in mines, my interest in geology was most stirred by fossils, and many a happy camping trip was spent in exploring prehistoric horizons. The stories told in coal, chalk, coral, agate, and amber are adventures in imagination. Registered in rocks are the details which tell how, when, and where; that reveal to us "what the first Morning wrote."

Nature works in cycles. The granite on high peaks crumbles away with raindrop and snowflake. Primroses and forget-me-nots bloom in high alpine meadows of granite soil scarcely finger deep. Snow-fed brooks carry the sand grains to the sea. There lime and other minerals cement the grains of sand into rock. In its turn, the sea bottom is thrust up into the sky and becomes a mountain summit. Sometimes this slow-dissolving sedimentary and storied rock may reveal its secret. The mind is stirred over the thoughts of horizons not yet risen in the landscape, over the higher evolved life that in other scenes and times will have its strange day.

Wet Spots in the Desert

A cluster of palms, flags of the desert, stood grandly near camp. In the background were a few stunted mesquite trees. A yucca held high its stalks of bloom, and somewhere nearby among the silky-yellow bloom and thorny lobes of prickly pear, a cactus wren had her nest. I was alone on the desert. The night was too balmy for a campfire and, happy and hatless, I walked round near my sleeping bag, looking at the stars, thickly sown in the strangely low-hung sky, and listening to each merry outburst of the desert coyotes.

A precipitous peak rose, barren and desolate, to the east of the camp. There was not a cloud in the sky, and the dry wash which came down from the heights did not seem to have felt water for years. When the air grew chilly, I laid my sleeping bag on the dry wash and crawled in, wondering if it ever rained on the desert. I had never heard of desert cloudbursts. Camping beneath a cracked and condemned reservoir dam would have been less perilous than the place I had selected for my sleeping bag.

I was awakened toward morning by a rumble, a slight trembling of the earth. It roared louder, and I made haste to get out of my sleeping bag, for if there was to be an earthquake exhibition in the desert I must see it. A black, broken wall of water was rolling down upon me from the peak slope above. Grabbing my outfit, I rushed for a nearby rock pile.

The water, with rush and roar, was sweeping over and round the rock pile as I climbed to the top of it. A deep black-yellow flood tore roaring by. The water rose knee-deep. Though infrequent, these brief local desert storms arrive with a rush, and burst with terrific violence. In a short time, the flood was gone and the sky was clear. The flood was mostly rocky debris. There was more solid than liquid. It was an earthy avalanche—a fluent landslide, mostly gravel, sand, and boulders. Water had given these lubrication and ball bearings. Liquid measure is not the medium for measuring desert floods.

Trainloads of debris were rushed down the slopes, and my camp ground was torn to pieces. One side of the old dry wash was ripped out to a depth of from eight to seventeen feet. My palm trees were uprooted and lay junked a quarter of a mile out on the flat desert. Trees, shrubs, and cactus were uprooted that had not been uprooted before. One uprooted mesquite tree had roots enough to equip a whole flotilla of long-armed octopi. Though less than sixteen feet high, its root system was ample for the ordinary grove. Desert plant life must have specialized roots. Its taproot, though a yard or so of the tip was broken off in the ground, was twenty-one feet long. I cut this off beneath the trunk, stood the tree up, and strung out its outreaching roots. These were from forty-one to seventy-three feet long. Had I been able to stand this mesquite in the pitcher's box, its radiating roots would have overrun all the bases and the long root over home plate would have reached beyond far enough to be an entangling snake for both the catcher and the umpire.

From my first camp I made a number of short excursions into the desert, keeping camp always in sight, so that I could reach it quickly in a forced retreat. But one day I bade farewell to every fear and with pack on back set off for a spring in the desert two day's walking distant.

Along the mountain edge of the desert there were gullies from down-rushing floods and deposits of gravel and debris flooded down from the mountains. One mile took me beyond deep gullies or arroyos out in level distances that had no horizon except the vague sky. The sage which I had known on the plains was stunted and only thinly scattered. But it had the good, pungent scent and the sage green which distances toned to purple. Occasionally, there was a prickly-pear garden with plants of giant size and numbers burning yellow candle blossoms. One giant cactus stood thirty feet high, with three short twisted arms halfway from the top. A jack rabbit hopped away from its shade. A little owl peeped from a woodpecker hole in the trunk. When I tapped the trunk with my staff, a woodpecker came from a hole in an arm. Other giants of the same species, the saguaro, were scattered about, a full stone's throw apart.

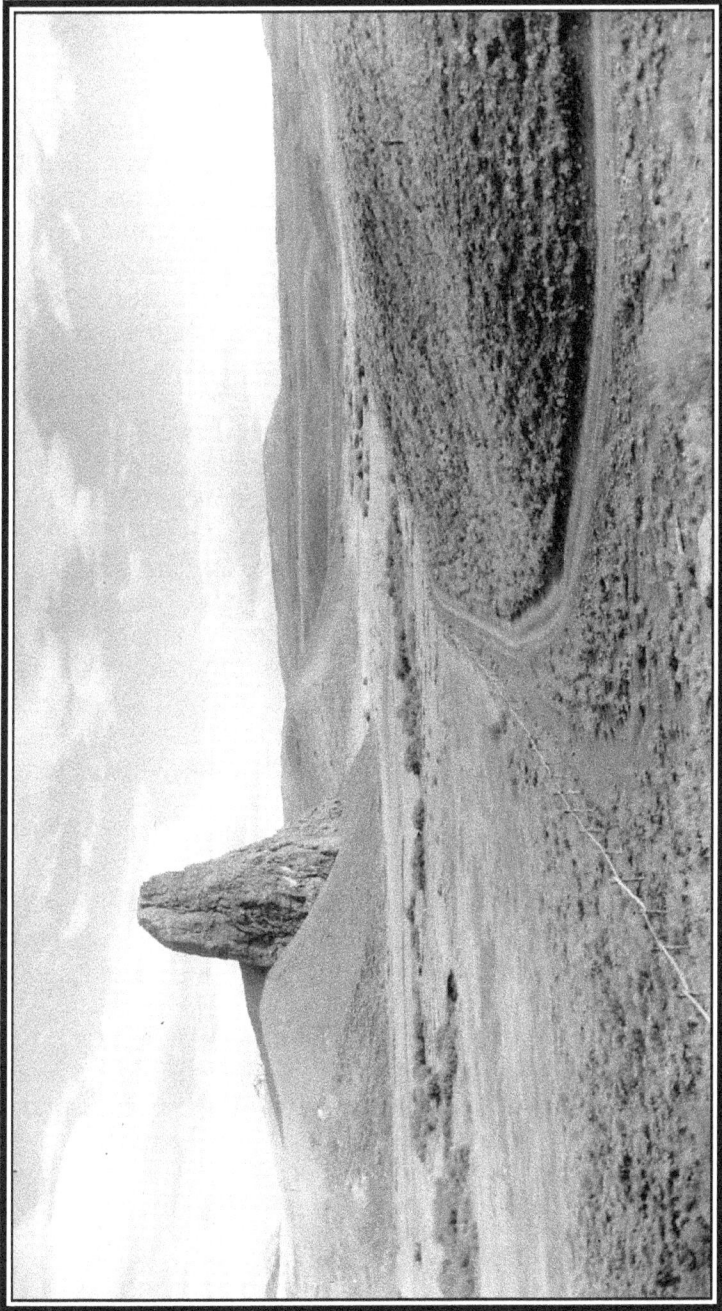

"Out Where the West Begins"

The nearest giant, fluted of form, decorated with sprays of thorns and spines, was a splendid set piece or lamp post of odd design and lighted with great white blossoms.

The wide spacing of most desert plants with barren earth between growths struck me as artificial and suggested nursery stock or experiment-station planting. Some distributions followed geometric lines. A species here and there had a tract by itself—still another artificial element. I had been at Burbank's a few days before, and carried fresh recollections of plants grouped and spaced. Nevertheless, the desert does give the feeling that it is artificially planted. And also that the planting was hardly a success, only one plant or seed of many having grown.

The giant cactus was as widely scattered as village lamp posts, while the tar-smelling creosote bushes were spaced about ten feet apart. Sage bushes, as a rule, were only a few feet apart.

All the first day I walked without crossing an arroyo or a sand drift of any size—just miles of level desert plains. The desert is extra dry. Dust, heat, and barren stretches, its travelers ever have with them. So, too, its drifted and drifting sands. Lack of water is something ever in mind. But there are oases in the desert—many of them. And there are trees, grass, and flowers, birds, animals, and beauty. While often, in later trips, I went out of my way to see the worst that deserts hold in their heart, most of the time within their strange borders I sought the best.

There is little need for raincoats or umbrellas. To the sun lover, the desert offers about ninety percent sunshine. And such air! It is as pure as is to be found on the earth. It exhibits mirages.

The first night, I made camp without water, but one canteen was full and the following night I should be at a spring in the heart of the desert. I camped by a silver-gray incense bush with starry flowers aburst in yellow glory. As I sat in the coming darkness, I noticed a few large eerie white spots showing in the sand. I had not heard of any white bird or animal. They did not move—they were white primroses.

The second day I crossed sand dunes high as hills and

miles long; dunes that travel and bury and smother the scanty vegetation in the front. Over the dry-as-dust dunes, the heat shimmered and sizzled.

I came upon a yucca standing upon a ten-foot stilt. Apparently, the sand bank had drifted it under and it had lengthened its stalk, extended its head, and risen above; again it was buried beneath the deepening sand. This time it lengthened its neck about three feet. This appeared to have been done three or four times. Then the dune, moving on, had left its head up in the air.

Late the second afternoon, my dim trail forked. There was absolutely nothing to indicate which fork led to the spring. As my canteen was nearly empty, I was concerned. There was nothing in the level distances or the desert plants to recommend that either trail would lead to water. I took the fork to the right, followed it a few minutes, doubled back, and took the other. The sun went down, and the sunset sky showed plains of smokeless embers and mountains of burning gold. Then twilight gave dimness to distances. Still no spring. A star rose over the desert. I came to scattered bones, then to the spring. There was only a trickle and this of alkaline flavor.

When I awakened in the morning, a mockingbird was singing. In the moist earth near the spring were a few square yards of salt grass and a few other plants. There were antelope and sheep tracks; nearby was a flock of desert quail. I filled my canteens and started for the next water hole, fourteen miles away according to the map.

There more often is eloquence than monotony in the level, limitless distance of the brown dry desert. Again and again, I turned to look across the long, long expanses to the hazy possible horizon. In the dry sands and unbroken sunshine I came upon an ocotillo shrub in bloom—an upright bundle of long thorny whips, each tipped with a short lash of scarlet flowers. It is one of the strangest of all the striking plants of the desert. Later, one of these blooming plants in a wind storm struck and lashed about with its flame-tipped whips in a manner almost uncanny. The rich, rare flowers of most cacti compel one to forget their fierce, thorny setting.

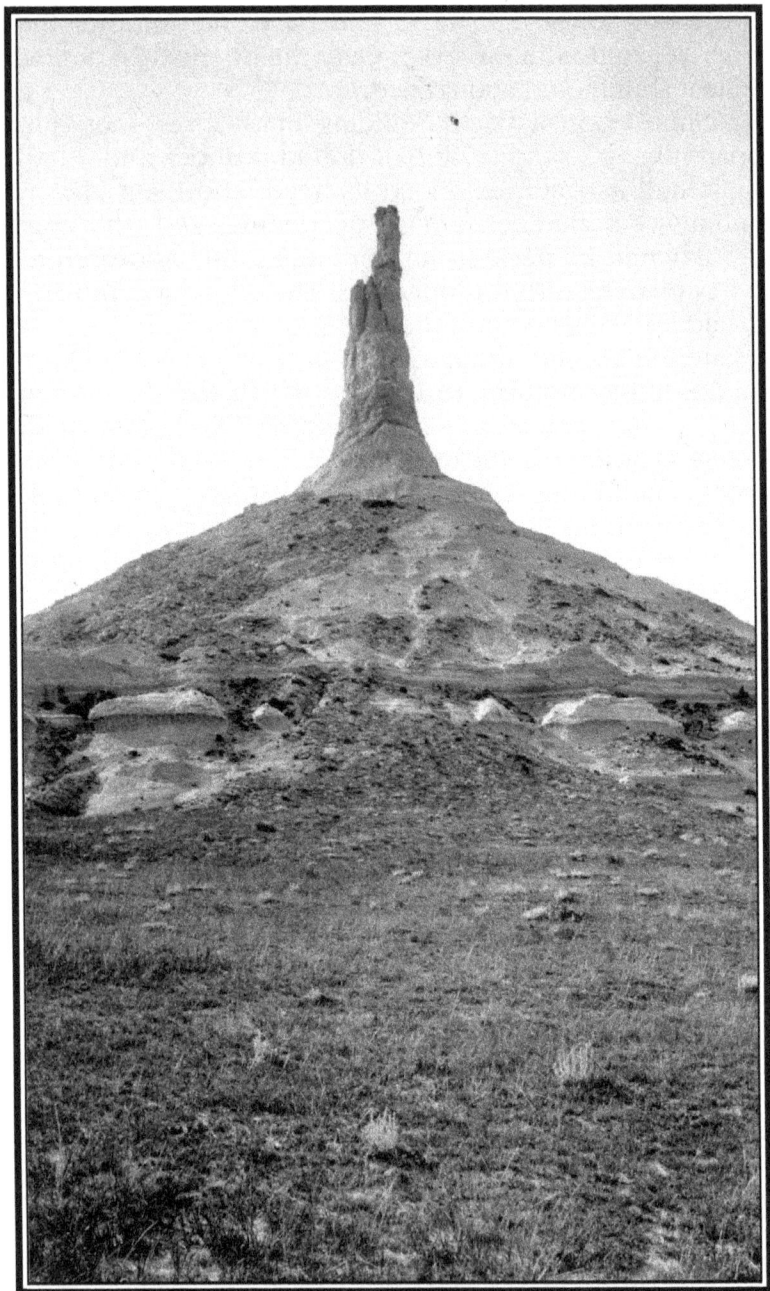

Chimney Rock, west of Camp Clarke.

Their rich blossoms burst from a wreath of thorns. The barrel cactus flares forth in greenish yellow; the strawberry cactus carries blossoms of ruddy claret color; and I passed several hedgehog cacti that were bristling thorns and robed in rose-purple flowers.

I passed within a hat's throw of the alleged water hole without seeing or smelling it. Late afternoon, I commenced to circle for it. Just after dark, I smelled it. One cannot choose a desert water supply. This one was impossible. A bloated snake and two dead rabbits afloat! I might have boiled the water, but I did not. With one canteen empty and the other growing light, I set off through the night for the nearest spring. Whenever I tried to get along with less than two gallons of water per day, it was too long between drinks, and I could think of little else than water. On a hot desert, a man may live thirty hours without water, but many have lost out in less time. The trail long used by wild folks, I followed easily through the night. At sunrise I came to the place, but not the spring. The place was dry. I was near to choking with thirst and alkali dust.

Exploring in every direction with my glass, there appeared to be black objects by a low bluff about two miles distant, probably barrel cacti. These small, green, thorn-covered barrels are filled with watery pulp. Each is a tiny oasis. I hurried toward them.

When almost to the little cactus barrels, a dust storm swept suddenly upon me. I could not see through the dust and sand that filled the air. It was smothering. I drained the last drop of water in my canteen and lay down, not to the leeward, but to the stormward of a giant cactus. Sand showered over me, but this side was less dusty than to the leeward. The dust devils sang and whirled round me. These storms sometimes last a day or longer and the air filled with fine light dust is dull and hazy for several days afterward. With a handkerchief over my nose I simply endured and allowed quarts of sand to sift in beneath my clothes.

Toward evening the wind stopped abruptly. I started for a barrel cactus, which was within two minute's walk. My tongue was swollen, and the alkaline dust in my nose and throat made breathing difficult. In my nervous haste to cut the top off a barrel cactus, I cruelly pierced and stung my hands, wrists, and legs with the thorny spines. But once the top was off, I battered the pulpy interior with my staff, then grabbed a handful of the juicy pulp. It had a turnipy taste, but it was refreshingly wet. I pounded pulp, sipped and ate it for an hour or longer, then camped for the night. After emptying two barrel cacti I started next morning across the desert for the mountains some miles from the place where I had been flooded out of camp.

On more than one desert trip, I have found the barrel cactus a life saver. It is a living storage tank of nonporous, dark green, rubbery, seamless material. It is well guarded with thorns and spines decoratively arranged. It is fluted with deep accordion pleating, and contracts as drained, thus exposing less surface to the suction of the extra dry air. I was sitting by one of these fellows during a rain. It speedily commenced to fill. I saw it begin to swell. It seemed to be moving. I stepped back; I did not want this thorny fellow to edge or roll against me. Was it unfolding or was I succumbing to a mental mirage of the desert? was my first thought.

"Endure long periods without water," is the first law of the desert. All desert plants are highly specialized in seizing water swiftly and storing it securely. Most leaves are hastily discarded when moisture time is over. Many of the cacti are leafless and are built so as to have least surface exposure. The stalks, twigs, and essential leaves are formed so as to have the least exposure—to put the greatest check on evaporation; and many leaves and stalks shellac and seal their exterior so as to prevent loss of water.

Numerous desert plants have long outreaching roots whose spread of spiderlike webs quickly catches and absorbs the precious water. Water is seldom delivered, and it sinks rapidly through the porous surface. An extensive water-getting system is a necessity. Many plants have enlarged trunks or tuberous growths for storage reservoirs. Those under the earth are well secured against hungry teeth and the ever-thirsty air. The trunk reservoirs are defended by thorns.

Desert plants must be scattered, widely spaced, that each

may have an extensive water-catchment space. Two or more species may occupy the same territory, but each probably has a different root system. One with finer roots may penetrate more deeply and thus catch the remnants missed by the species with the uppermost and larger roots.

Many trips into the Arizona, Nevada, and California deserts impressed me with the special adaptations and adjustments of desert animals and plants which enabled them to succeed with so little water.

Geologists say that all life was born in the sea, and that there it lived for almost an eternity before it even ventured as far as the moist and slimy shore. Plants were thus started and had long development in the water. Desert flora has traveled farther on evolution's changing way than any other flora. It is highly developed—intensely specialized. Most leaves have been discarded, and in meeting the requirements of the desert they have developed non-leak and guarded storage tanks, thorns, an intricate far-reaching root system, and other specialities that enable these plants triumphantly to cast a shadow on the desert's dusty face.

On this trip, I made the acquaintance of the cholla cactus, perhaps the most stinging, insidious, and cursed of all thorn bearers. Its small thorns are shed annually and are scattered by the wind. On the stalk or the earth these thorns pierce clothes and shoes. I have sometimes lain down upon them, and have fallen upon them or brushed against them while hurrying to follow bird or coyote.

The desert is the producer of thorns and spines. These probably are an evolutionary development of leaves and stems. Thorns are microscopic and gigantic, big as bayonets; they grow singly, in clusters, and in graduated groups; they are hooked, curved, and sometimes poisonous. The desert is fierce.

In traveling without a gun on the desert, I found a little wild food—Indian fig—the fruit of cactus, chia, and the beanlike seeds of the mesquite tree; the pulpy interior of some cacti, all of which reminded me of the raw roots; the fleshy and semi-tuberous reservoir roots of a few plants, wild apricots, the fruit of palms, young shoots of yucca, the inner bark of trees; and pinion nuts in border zones of deserts.

The desert has soft color, gentleness, and beauty. There are humming birds, butterflies, bees, and wild flowers. There is interdependence as well as individuality. There is mutual aid. Most desert flowers are pollinated by their winged friends. The intimate interrelation of the pronuba moth and the yucca is one of the strangely fascinating stories.

All day I traveled with the distances hazy from the dust storm of the day before. After a prolonged and high wind, two days or longer may pass before the lighter dust particles settle from a desert sky.

I reached the mountains and made camp, being careful to keep away from the slope of a canyon which might any time send down an avalanche flood. The spring had been wet and the desert had responded in a floral surprise, an exhibition of desert wild flowers. The wondrous wildflower garden on Mount Rainier did not impress me more than these strange wildflower gardens of the desert. I saw a mile-long ragged foothill garden of pink verbenas set off with rock piles, leafless cacti, and heavy-topped palms. There are a few hundred kinds of wild blooms on the American deserts.

Bloom time of desert wild flowers commonly is short. When conditions are normal, they bloom at a normal time. But dry springs delay or even prevent the blooming of many. In places, plants and seeds may wait years, then comes a rain, and they burst into blossom. Among the brighter plants seen on the California deserts were yellow encelia, dark-red beloperone, and blue phacelia. These are handsome, but the wild red hollyhock was for me the poem of the desert.

Three palms appeared to be looking into a canyon near camp, and I moved over for a look. It is a wonder the stiff cacti did not also try to look in. Down the north facing wall of the canyon there appeared to be pouring a colored cascade of flowers. Red, yellow, white, and blue blossoms enriched the wall up to where it seemed to touch the sky. A few days later, I returned, and the grim rocks, with only a few splashes of green, denied wearing the wondrous wardrobe. It had vanished like a mirage.

Plants store water and ration themselves during the numerous dry days. But I cannot understand how desert animals get along with so little to drink. Antelope go for days without water; so, too, do rabbits, skunks, rats, and gophers. This coyote can go long periods without drinking, but he is a rapid traveler and appears to visit the water holes more frequently than other large desert animals. A bird or animal may eat food more or less filled with watery tissues—blood for flesh eaters, pulpy tissue for some, and juicy plants for others. Sometimes all the food of all these is almost without moisture. How birds and animals go without drinking for days and even weeks is one of the unwritten desert stories.

With canteen water and cactus pulp, I spent four days in Arizona on the summit of a desert mesa watching a flock of bighorn sheep. They ranged about eating dry vegetation. Not once did they go down for water, nor did they have any from pockets in the rocks.

When I last saw them they lay resting contentedly. Apparently, they were not thinking about water and had no interest in the possibilities of rain.

It is strangely interesting that about one fifth of the earth's surface in each of the past ages should have been desert—and that about one fifth is desert today.

Books have infested the desert with gila monsters, hydrophobia skunks, tarantulas, side-wheel rattlers—fellows that move as though cross-eyed, cross-geared, and strike without buzzing. Just how deadly any of these fellows are I cannot say; they kept so well out of my way and I saw so little of them that little was learned of their biography. My species had terrorized them.

The desert is the land of color. Color is somewhere all the time. Wild flowers give it a brief though brilliant day. Sage gives leagues and rolling seas of gray or purple; there are graduated level distances of golden brown, tawny gray. Cloud shadows, cliff shadows, and mirages are likely to be of purple and may be broken with silver and backed by gray or gold. Shadow continents of dusty rose and islands of dusty yellow often make cloud scenery in the sky; there are sunrise campfire horizons and clouds of molten opal or drifted snow edged with copper for sunsets.

The eye constantly underestimates desert distances. "Telescopic air," was a Mark Twain expression which I did not appreciate until I looked through distance-concealing desert air. I had thought my eye range finding faculty was accurately developed. Twice I seriously misjudged distances—taking an entire day to travel to an object the time distance to which I had judged would be two hours. It is difficult to develop and adjust an eye to the perspective of the desert.

One night, I arrived long after dark at a water hole called Mesquite Springs and found a man by a mesquite campfire. By the fire he told me of a desert adventure—the result of a distance misjudged. Without a canteen, he had set off for a palm cluster by which there was a spring. He would be there in an easy hour, as the palms were in sight across the desert. For more than two hours he walked with the palms in sight, and on their disappearing behind a foothill, he turned back. After walking for more than three hours, he felt that he had missed the way to his former campsite. In need of water, he concluded to try again for the spring by the palms, as the palms were again in sight. The place which he had left in the morning was within a stone's throw when he turned. The day was hot, his thirst terrible. After walking nearly three hours toward the palms, he again lost sight of them. He was lost. He had counter-marched and did not wander here and there as so many do on the desert. But he was muttering, throwing his clothes away, and chasing something in his delirium when a prospector from the palms came out and rescued him.

Desert moonlight is the most moony effect that I have felt. It eloquently emphasizes the strangeness and the barrenness. In it one readily imagines himself not on the earth but on the moon. The cactus world is weird; the scattered sages and sand dunes make another strange land. A wind makes the earth infinitely old.

Never have I felt more primitive than one moonlight night on a barren desert island in the Gulf of California. I was alone, miles from anyone. Out of food and fresh water,

I thought to find something on a little island offshore. I pushed out on a piece of wreckage. The water seemed unreal. The desert mountains rising from the sea were bold and barren beneath the moon. On the shore, I found oysters. As I sat by a giant cactus of primitive prehistoric form, roasting oysters in the little fire, the world had no interest in me, and in it I was a mystified primitive explorer.

THE END

The story of geology – of the earth
beneath our feet – is so fascinating
that if commonly known this
ample and illustrated field would
appeal to the leisure hours and
delight

A handwritten note of Enos' on the story of geology.

Enos A. Mills

They carved for him a granite grave,
    From out the recesses of nature's wonderland,
Where years ago the luring call of loneliness
    Had whispered strains of hope to roving Indian bands
He cared not for the glitter, or the falseness,
    Of the artificial wealth within the world;
He understood the song the river sang,
    The bold defiance the mountain lion hurled
A challenge to the human flood to come,
    As the answering hills confirming echoes rang.

He asked so little from the world of men,
    But deeply drank of nature's lasting charms;
He gave so much to those who understand
    The soothing lullaby, the encircling arms
Of Mountains reaching lazily toward the sky;
    Of canyons, winding serpent-like thru towering hills;
Of roaring cataracts a-thundering toward the sea,
    Yet pausing long enough to kill the little rills
That feed the parent streamlet, while it carries on,
    And sings its song of joy to you and me.

The ghost like peaks in silence, guarded well,
    For years, the dreams within the hermit soul
Of the inmate of that lonely mountain cabin,
    Until one day this vision claimed its toll,
And worked upon the heart strings of this man,
    And looking far ahead into the coming years,
He sought to share his fairy land with men—
    And build within the vastness of this land a peer
To all the other play grounds in this favored land,
    Where towering mountains guard the entrance to the
    glen.

Unlike so many dreamers who have lost
    Their battle for the future of the race,
He lived to see his fondest hopes fulfilled;
    To see the throngs from every spot and place
Within the confines of this mighty land of ours,
    Come journeying as pilgrims did of old;
Not seeking to pollute his land of dreams—
    No fighting, struggling mass—just craving gold.
They builded well, who made his resting place,
    Where the moon will always cast its kindliest beams.

Pierce Egan, Loveland

GEOLOGICAL LABORATORY
JOHNS HOPKINS UNIVERSITY,
BALTIMORE, MD.

Feb. 20, 1905.

Mr. Enos A. Mills,

    Estes Park, Colorado.

Dear Sir:

    Thank you very much for your letter of Feb. 3rd with
the photographs of the Hallett glacier.   The rate of motion
which you have obtained does not seem to me to be remarkable.
The means which you used were not, of course, very accurate, still
I believe that you have made a fair approximate determination
of the rate of motion.   Your idea of making various photographs
of the glacier is a good one and I suggest that you make photographs from two stations on opposite sides of the glacier at a
little distance from it though at a higher altitude so as to show
very clearly the position of the end of the ice with reference
to surrounding objects.   Photographs from these stations repeated
at the end of successive summers will show very clearly the changes in length of the glacier.   A small sketch map of the glacier itself and the surrounding mountains would also be very use-
ful to show the slope of the ice and the valley of the glacier.

    I am sending you a few of my earlier reports.

                    Yours very truly,

                    Harry Fielding Reid

224

JUNIUS HENDERSON
JUDGE

Also Geologist

BERTHA M. THOMPSON
CLERK

COUNTY COURT
BOULDER COUNTY
COLORADO

BOULDER, COLO., January 17 190 8

My Dear Mr. Mills:

I was very much interested in your brief account of your recent trip to Hallet Glacier and the two views you sent. The rate of movement you mention (1.1 inches per day), is rather unusual for a glacier of that size. It equals about 33 feet per annum. The rate in the winter would probably be less and in the summer more. Please let me know how the measurements were made and whether you left any tablets and bench marks so that the annual rate could be determined next fall.

I wrote to Prof. Harry Fielding Reid, American member of the International Glacial Record Committee, about your measurements, so you need not be surprised if you receive a letter from him.

With sincere regards,

Yours Very Truly,

Junius Henderson

225

COUNTY COURT
BOULDER COUNTY
COLORADO

BOULDER, COLO., *Sept. 19* 1905.

Mr. Enos Mills,
 Ester Park,
  Colo.

My Dear Mr. Mills:

I am anxious to learn something about the Sprague glacier, near Estes Park, particularly as to its size, etc., compared with the Hallett. Can you give me the information. As you know, the Hallett is quite broad but very short, having melted back practically to the bergschrund, where the ice of a glacier proper through accelerated movement breaks away from the nivé, hence its title to the name "glacier" is sometimes questioned. I am anxious to learn how the Sprague compares. Any information on the subject will be most welcome. Our observations on the Arapahoe show that Station No 1 300 feet from the north edge of the ice moved 11.15 feet and Station No. 10, 1000 feet from the edge moved 27.7 feet from Aug. 30, 1904 to Aug 30, 1905.

Yours very truly,
Junius Henderson

Estes Park, Colo. Sept. 23, 1905

Mr Junius Henderson,

Boulder, Colo.

My Dear Mr Henderson; - Yesturday I had an hour at the Hallett glacier but had time for a few pictures only. I regret that I cannot get time to make any measurements this fall. However, the glacier is a trifle larger than last September. Will mail you some pictures later.

As to the Sprague glacier; I was there ten years ago and estimated its height at eleven hundred feet, and its width at fourteen hundred. Its slope is much steeper than that of Hallett glacier. It is situated on the northerly slope of Stone's peak about twelve miles west of Estes Park Postoffice . My barometer showed its bottom be just twelve thousand feet above sea level.

If I can get hold of a photograph of this glacier will mail it to you.

I thank you for measurements on the Arapahoe glacier.

very truly,

Enos' reply to Junius Henderson, judge and geologist.

Ironside, Ore.
Jan 31-20.

Mr. Enos A. Mills,
Estes Park, Colo.

Dear Sir:-

Your article, in a recent issue of The Sat. E. Post
"Sight-seeing by Wireless", interests me. I suppose that we
are most interested in those subjects with which we are more
or less familiar?

I have spent years in the land of the mirage- New Mexico,
Arizona, Lower Calif. and the deserts of the Mojave and the
Colorado and am quite familiar with nearly all of the features
you write of. You have gone my experience one better, however,
in seeing the bear around the corner. All of the visions
I have noted have been reflected directly above, in the
strata of air over the object.

This does not so much apply to the distorted pictures we
often think we see, such as palm trees where there is not
even short grass to be reflected. I have often seen islands
in the Pacific, when I was so far that they were away below the
horizon, but they were always directly above their true position
as proved by my taking a compass course and sailing to them.

Often a second island is reflected above the real land and
resting on top. This is not confined to the warm waters, for
I have seen it in the Arctic, when the mercury was 40° below.

In this same connection, I will say that I have seen the
" heat waves" that we so often see over the hot sands of the
south, just as pronounced over the ice in the north and at
such times the same effect is noted. Icebergs are distorted,
They often stand up on stems, like mushrooms· If anyone
thinks heat is used in connection, I have only to say, 'go see'.
It was from 40 to 50 below when such conditions were met with.

Another feature of the north, tho not strictly a mirage.
Frequently there is an absolute lack of horizon, the snow
and sky blending so that one can not tell where either begins
or ends. With absolutely nothing to break the monotony there is
no prospective. It is out of the question to judge distance.

I was once tramping over the trail, well to the north of
any timber, when I sighted two large pines, on a ridge about
five miles away, so I judged them to be. I knew something was
wrong, for there were no trees, of any kind there, but I
headed that way, to find out. Perhaps one hundred and fifty
feet was covered, when I came to two grass stems, about a foot
above the surface of the snow. The ridge they were on was
a muskeg. Such things happen every day, when the sky is over
cast, to confuse the horizon.

I have always been more than interested in you papers on
the beaver, especially one you wrote some ten years ago, which
was published in the Post. I felt that I knew, personally,
every one of them. It took me back to the middle of the 70s,
when a bare footed boy ranged about, from Springdale, Jimtown
and Estes Park down to Boulder. Later he wandered as far as
the Southwestern corner of the state and knew about every

rock and stump.

I have a beaver problem on my hands, here. This settlement is at the base of the Blue Mts., of Eastern Oregon. and we have a fine little stream, well stocked with beaver. There is a dam for every four hundred yards, for at least fifteen miles' One dam within two hundred yards of the postoffice.

Down below, R.N. Stanfield has a ranch. He poses as the largest sheep onwer on earth and is backed by the Swift Co.

He is now trying to get a permit, from the Gov. to kill all of the beaver and tear out the dams, claiming that he is deprived of water, that he would get if he was allowed to kill the beaver.

Some twenty years ago, there was a halfbreed trapper came here and talked the ranchers into permitting him to trap the beaver, on a commission, the ranchers to get half the fur.

He did a fair job of it, and by spring was supposed to have the last one. He went out to sell the fur, but neglected to either come back or to send the ranchers any of the profits.

within two years the creek had washed out most of the dams and scoured down to the underlying sand strata and in place of a slow running creek, with banks full, they had a dry, deep wash. Ranches that had been cutting a hundred tons of hay stacked up eight or ten tons. Then they saw the holes in the ladder. As usually happens in such cases, there were a few beaver left and by degrees they began to rebuild the dams. They have, now become as plenty as ever and are, rightly regarded as the best friends we have. Those who have then on their ranches, say they are worth at least $1000 to each ranch. There will be something of a fight, before anyone kills them but Stanfield is a strong man, both as to his money and pull in politics and he will be a hard one to down.

One more on the beaver, and Ill let you rest.

One of my neighbors was trapping for muskrats and caught a halfgrown beaver. Tho the trap was small, it had the beaver just right to hold him, without doing any harm. Not wishing to injure it, my friend took a stick and carefully pressed down on the spring, until the foot was free. Naturally he expected the animal to plunge into the creek and be gone, but it walked down to the edge of the water, paused and looked back, then came back and smelled about the feet and clothing of the trapper, looked up into his face a while and then calmly swam off, on the surface and around the bend of the creek.

As a matter of fact, they are about half domesticated

Yours sincerely,

H. V. Armstrong

September 27, 1922.

Mrs. Enos A. Mills,
Long's Peak, Colorado.

My dear Mrs. Mills:

I never had the pleasure of meeting you, but I have had the great good fortune of knowing your husband some twenty years. I invited him to make his first address before a student body when I was professor in the University of Denver and I have followed his career since then. He spoke in our Assembly here in El Paso last winter and carried the students away with him. Monday in Assembly a memorial service was held in his honor. The announcement of his death greatly affected the student body, who remained silent in recognition of your husband's services to the world.

If it is convenient, I would appreciate it very much if you will send us the titles of the three poems read at his funeral. I would like to have them read in our Assembly of 1500 students.

Mrs. Roberts joins me in the deepest sympathy for you in your dark hour of trial and tribulation, but there is this glorious thought -- that your husband's immortality is certain.

I am

Yours in high regard,

FHHR:SDW

Frank H H Roberts

Loveland, Colorado, Oct. 3, 1922

My dear Mrs Mills,

My wife and I were pained on returning from a several weeks stay in the mountains beyond reach of the mails to learn of Mr. Mills death.

We, together with his thousands of other admirers mourn with you the passing of so rare a spirit.

I am greatly disappointed that with all that is proposed in honoring his name and "carrying on" his educational work to have seen nothing as yet about aiding you in "carrying on" his fight against the monoplies.

I have prepared an article about this that I shall try to get published soon.

Sincerely Yours

Eugene Smith

Country Life

GARDEN CITY   NEW YORK

October 19th, 1922.

Dear Mrs. Mills:

Thank you for your letter of
October 14th. Mr. Mills, just shortly
before his death, sent me a letter about
making the change in his story, "News from a
Fossil Lake", which I have incorporated in the
manuscript.

I cannot tell you how terribly
shocked and grieved I was on my return from
a month's trip in the woods (where of course I
got no news) to find that Mr. Mills had died
suddenly during my absence. Our relations,
chiefly through letters, were always so friendly
and he stood for such splendid ideals and principles
that I cannot tell you how badly I feel that
he has gone. Although I had only the pleasure of
seeing him personally once, I felt I had in him a
real friend and I had hoped some day to see something of him out at Long's Peak. I sympathize with
you most deeply in your great loss, and if I can
ever be of any assistance in any way, please do not
hesitate to call upon me.

Sincerely yours,

Reginald T. Townsend

Editor.

Mrs. Enos A. Mills,
Long's Peak, Colorado.

100M—5-22   Form 1734

MISSOURI, KANSAS & TEXAS RAILWAY
THE MISSOURI, KANSAS & TEXAS RAILWAY OF TEXAS
WICHITA FALLS & NORTHWESTERN RAILWAY
C. E. SCHAFF, Receiver

San Marcos Texas Oct 14th 1922.

Mrs Enos B.Mills,
        Denver Colo.
Dear Madam:-

      I trust that you will pardon me for this seeming intrusion,but
I have just learned of the death of your Husband and my dear friend,Enos
B.Mills,this was indeed a great shock to me,but I know it cannot be com
pared with the great loss of your Husband.

      While I never had the pleasure of personally meeting Mr Mills,I have
read so many of his writings that I felt I knew him,in fact he became my
big brother,he has helped me in so many ways,he has made my trails
easier,and my burdens brighter,I like him,enjoyed trips alone, miles upon
miles I have travelled in the hills of Southwest Texas,and almost every step
I have swon Mr.Mills.

      He must have been Godly man,he was continually out in Gods
great open, where God showed him His wonderous works and His plan of creation
but Mrs Mills have you ever thought,(no doubt you have) what wonderous
sights he is now enjoying,he has no shadows now, all sunshine,all the rough
places are made smooth, all the crooked places made straight,I envy his
present estate,and I am more determined to better my ways that when God
calls me west,to the river that bounds the unknown shore I may have my pack
already,I believe we shall knew each other,in the great beyond.

      I want to come to Colorado in the spring,in fact am now laying
my plans in that direction,and if I am permitted I surely shall make my
way to Longs Peak where I learn that all that is eartly of Mr Mills is layed
there around the place he loved so well while he was on this earth,

      I wish it were possible that I could be in the position to
give you comfort in this hour of grief,I can only point you to the great
God and Ruler of the uni verse who is ready and willing at all times to
help all those who earnestly seek Him.

      I respectfully beg to remain.

      Sincerely yours,to command.

Walter F. Hall

           Agent,M.K.&.T.Ry

      San Marcos Texas POBox 65